Bithia Mary Croker

A Bird of Passage

Vol. 2

Bithia Mary Croker

A Bird of Passage
Vol. 2

ISBN/EAN: 9783337813567

Printed in Europe, USA, Canada, Australia, Japan

Cover: Foto ©berggeist007 / pixelio.de

More available books at **www.hansebooks.com**

A

BIRD OF PASSAGE.

BY

B. M. CROKER,

AUTHOR OF "PROPER PRIDE," "PRETTY MISS NEVILLE,"
" SOME ONE ELSE."

IN THREE VOLUMES.

VOL. II.

" Such wind as scatters young men thro' the world
To seek their fortunes further than at home,
Where small experience grows."

THE TEMPEST.

SECOND EDITION.

London:

SAMPSON LOW, MARSTON, SEARLE, & RIVINGTON,

. CROWN BUILDINGS, 188, FLEET STREET.

1887.

CONTENTS OF VOL. II.

CHAPTER I

"Mr. Lisle has given me a Ring!". . . . 1

PAGE

CHAPTER II.

"Why not?" 21

CHAPTER III.

"Stolen from the Sea!" 27

CHAPTER IV.

The Ball 47

CHAPTER V.

"But what will Papa say?" . . . 65

CHAPTER VI.

Proof Positive 91

CHAPTER VII.

"A Great Battle" 104

PAGE

CHAPTER VIII.

The Nicobars 121

CHAPTER IX.

The First Grave 137

CHAPTER X.

"Was it Possible!" 146

CHAPTER XI.

"Farewell, Port Blair" 168

CHAPTER XII.

The Steerage Passenger 184

CHAPTER XIII.

A Poor Relation 200

CHAPTER XIV.

In which Everything is settled to Mrs. Platt's
Satisfaction 220

A BIRD OF PASSAGE.

CHAPTER I.

"MR. LISLE HAS GIVEN ME A RING!"

"Vouchsafe to wear this ring."
Richard III.

FOR several days after this startling occur-
rence, Miss Denis did not appear in public.
She would gladly have denied herself to
all visitors save Mrs. Home; but who could
shut out Mrs. Creery? She penetrated to
Helen's room, and from thence issued daily
bulletins to the whole station in this style:—

"The girl was knocked up, her nerves
were unstrung. She was in a very weak
state. She required rousing!"

Miss Caggett also forced her way in, and
imparted to her friends and acquaintances
"that, from what she saw of the invalid it would
never surprise *her* to hear, that there was

VOL. II.   B

insanity in the Denis family, and SHE would
not be astonished if she was going off her
head!"

This affair had given Mrs. Creery something
fresh to talk about, and she related the whole
story at least thrice separately to every one in
Ross, and as often as she had opportunity to the
people from the out-stations.   On each occasion, she added a little touch here, and detail
there, till by the end of a week it was as
thrilling a narrative as any one would wish
to hear.   Mrs. Creery flattered herself that
she told a story uncommonly well; so also
said public opinion,—but then their reading
of the word *story*, was not exactly the same
as hers.   She had brought herself to believe,
that she had been the only person on the
wreck who had evinced any presence of mind,
and it would take very little to persuade her,
that she herself had been in personal conflict
with Aboo—Aboo who had been duly hanged,
at Viper on the succeeding Monday morning!
She now commenced all conversations with,—

"Of course you have heard of my terrible
adventure on the wreck? and the marvellous

escape we all had?" and then, before she could be interrupted, the rehearsal was in full swing. This intrepid, loquacious lady, entirely ignored Mr. Lisle,—of whom Dr. Malone reported, that he was nearly convalescent, the cuts from Aboo's knife were healing rapidly, and that he was going about as usual at Aberdeen.

Mr. Lisle was among Helen's first visitors; and he came alone. He wore his arm in a sling—this gave him quite an interesting aspect,—and carried a small parcel in his hand. He was struck, as he entered the drawing-room, with Miss Denis's altered appearance; her face was thin and white, and her eyes had a startled, sunken look. They shook hands in silence, and for quite a moment neither of them spoke. At last he said,—

"I hope you are all right again?"

"Yes, thank you. And your arm?"

"Is well; this sling is only Malone's humbug. I have heard of you daily from him—our mutual medical attendant, you know—and would have been over before, only

he said you saw no one.   I have brought you
this."

"What is it?   Oh, my sketch!"

"Yes, I fetched it from the wreck.   I
thought you might not like to lose it."

"Oh, I don't care!   I had forgotten it.
But how *could* you go back to that horrible
place?" and she shuddered visibly.

"Why not?"

She did not answer this question, but said
in a rather husky voice,—

"Mr. Lisle, you remember what you said
to papa.   That was absurd.   Only for you I
would not be sitting here now.   No," raising
her hand with a deprecatory gesture as she
saw that he was about to speak, "if you had
not come that time, I know in another moment
I would have been dead."

"Was it so bad as all that?   Well, but,
Miss Denis, that I should drag that fellow
off was a matter of course—that's understood.
Do you think any man would stand by and
see that brute throttle a girl before his face?
But that you should interfere in my behalf,
was quite a different affair—you know that!

My life hung on a thread—I believe I was within ten seconds of eternity. If you had not made that dash when you did, I should have been a dead man. I owe my life to your courage."

"Courage! Oh, if you only knew how little I deserve the word! You would not believe what a miserable coward I am. I actually tremble in the dark; I dread to open a door—much less to look round a corner; in every shadow I seem to see *Aboo's face*. I never, never could have believed, that in so short a time I should have sunk to such an abject condition."

"You will get over it all right. It is the reaction. You will soon forget it all," he answered reassuringly.

"I wish I could—all but your share in it, I shall never forget that!"

"Miss Denis," he answered gravely, "I am not good at making speeches, like—" he was going to add Quentin, but substituted— "other people; but whatever I say, I mean. I shall always remember that you stood by me at a great crisis, just as a man might have done. If you were a man, I would ask you to be my friend for life—and I am not a fellow of many friends—but as it is—" and he hesitated.

As it was, she was the only girl he had ever cared two straws about, and she was in love with James Quentin.

" But as it is," she repeated, surprised at this sudden pause, " I shall be very glad to be your friend all the same." Then, with a sudden pang of apprehension lest she had been over-bold, she blushed crimson, and came to a full stop.

" Agreed, Miss Denis. If you ever want a friend—I speak in the fullest sense of the word—remember our bargain, and that you have one in me."

The conversation had become so extremely personal that Helen was glad to change it rather abruptly by saying,—

"I have something here belonging to you," opening her work-basket as she spoke, and carefully unfolding from some tissue-paper the ring from the wreck.

He received it from her in silence, turned

it over several times in the palm of his hand, and seemed to waver about something. At last he said with an evident effort,—

"Would you think me very presumptuous if I asked you to keep it?"

The young lady looked at him with startled eyes and vivid colour.

What did he mean?

Observing her bewilderment, he added quickly,—

"Only as a memento of last Thursday— not to recall the whole hateful business, but just to remind you," and he stammered— "of—a friend."

"I should like to have it, thank you; and I shall always keep it," she replied, "and value it very much. Papa!" to her father, who had just entered the room, "look here —Mr. Lisle has given me a ring!"

Colonel Denis started visibly, and was not unnaturally a good deal amazed at this somewhat suggestive announcement. He liked Lisle far better than Quentin. Despite of the latter's fascinating manners to most, he scarcely noticed Colonel Denis during his

constant visits ; he considered him a slow old
buffer, left him to walk behind, elbowed
him out of conversations, and altogether
folded him up, and put him by. Helen's
parent was an easy-going gentleman, but he
had his feelings, and he did not care for
Apollo, and he liked his pauper-friend Lisle.
Nevertheless, he was not prepared to give
him Helen—indeed, he had never dreamt of
him as being one of her cloud of admirers,
and he looked very blank indeed, to hear
his daughter say, " Mr. Lisle has given
me a ring ! " and saying it with such supreme *sang-froid,* as if it were a matter of
course !

Mr. Lisle read his host's face like a
book, and saw that, for once in his life, he
was quite capable of uttering the word
" No."

" It is only a queer old ring that I found
on the wreck," he hastened to explain. " It
fell out from behind the wainscoting in the
cabin, and your daughter was looking at it,
and in the subsequent confusion carried it
away. She wished to restore it to me now,

but I have been asking her to do me the honour of keeping it, as—"

"Certainly, certainly," interrupted the elder gentleman, greatly relieved; "and so she shall, so she shall."

"It just fits me, papa," she said, slipping it on her third finger, and holding it up for approval.

The two men gazed at it in silence, and made no verbal remark, but the same thought occurred to both—assuredly that strange old ring had never graced a prettier hand!

When Mr. Lisle had taken his departure, Colonel Denis said to his daughter, as he picked up the *Pioneer*,—

"I like that fellow—uncommonly; there is no nonsense about *him*."

"So you should, papa, if you put any value on me."

"That is a thing apart, my dear. But I had always a fancy for Lisle, for he reminds me of a very old friend of mine, who was killed in the Mutiny. His name was not Lisle, but Redmond; but, all the same, the

likeness is something extraordinary, especially about the eyes—and Lisle has his very
laugh ! ''

" Which you do not often hear," remarked his daughter. " I'm sure Mr. Lisle
is a gentleman by birth,—no matter what
Mrs. Creery says."

" What does she say ? "

" That she is sure .his mother was a
Portuguese half-caste from Chittagong."

" She be blessed ! " angrily. " Lisle may
have empty pockets, but he has good blood
in his veins."

" Mrs. Creery also says she notices—"

" She notices everything ! If any one has
a button off their glove, she proclaims it on
the house-top," rattling his paper irritably.

" I declare, papa ! " pausing in the act of
rubbing up the ring with her handkerchief.
" What do you think is in this ring ? "

" A finger, of course," without lifting his
head.

" No, you dear, silly old gentleman, but a
motto, and I believe I can make it out.
Listen to this."

Colonel Denis looked over his paper, now all attention.

"It is rather faint, but," holding it close to her eye, "the first is a big L. Love—me —Love me—and leave—"

"Love me and leave!" cried her father. "A pretty motto, truly! I could do better than that myself!"

"Wait, here's another word. Now I have it; here it is, 'Love me, and leave me not.'"

"Show it!" holding out his hand. "It's one of those old Posey rings. Yes, there is a motto, but it was not intended for you, my young lady—"

"Of course not, papa," colouring. "Mr. Lisle did not even see it." (We would not be so sure of that.)

"I could not make out what you meant, Nell, when you told me so suddenly that he had given you a ring.—I declare, I fancied for a second that—that—but of course it was utter nonsense,—and, of all people, LISLE!"

CHAPTER II.

" WHY NOT ? "

> " Friendship is constant in all things, save in the office
> and affairs of love."
>
> *Shakespeare.*

THINGS went on much as usual after this at
Port Blair; there were no more tragedies,
nothing startling to record, and people had
quietly settled themselves down to wonder
if Lizzie Caggett would catch Dr. Malone,
and when the Quentin and Denis engagement
would be given out ?

There had been the ordinary settlement
amusements, including a grand picnic to
Mount Harriet (the last place Lord Mayo
visited before he was stabbed on the pier
below). Mount Harriet was a very high hill,
covered with trees and dense jungle, and on
the top of it was situated the General's
country bungalow. He did not often live

there himself, but it was in constant demand by people who " wanted a change," also for honeymoons and picnics. From the summit of the hill, there was a magnificent view of inland winding water, islands, mountains, and sea; but this view was only to be obtained by a steady two-mile climb from the pier, and an elephant, Jampanees (men carrying chairs), and two ponies, awaited the picnic party.

The elephant at Mount Harriet was a character; he was fifty years of age, and his name was "Chootie;" once upon a time, he had got tired of drawing timber, and slaving for the Indian Government, and had coolly taken a holiday and gone off into the bush, where he had remained for three whole years. However, here he was, caught and once more in harness, waiting very discontentedly at the foot of the hill, with a structure on his back resembling an Irish Jarvey, minus wheels, which was destined to carry six passengers.

Helen, and Lizzie Caggett, with happy Dr. Malone between them, went on one side; Mrs. Creery, Mr. Quentin, and Mrs.

Home on the other, and presently they started off at quite a brisk pace; but the day was hot, the hill-road was rugged, and "Chootie" paused, like a human being, and seemed to express a wish to contemplate the landscape. His mahout expostulated in the strongest language (Hindustani). "What did he want?—water? Then he was not going to get water—pig that he was!" Nevertheless he exhausted his vocabulary in vain.—Vainly did he revile Chootie's ancestors in libellous terms; Chootie remained inflexible, until two policemen armed with very stout sticks, arrived and whacked him with might and main, and once more he started off again, and kept up a promising walk for nearly half a mile; and now the praises lavished on him, by his doating driver, were even sweeter then new honey; but, alas! he was praised too soon. Without the slightest warning, he suddenly plunged off the road down a place as steep as the side of, not a house—but a church; deaf to Mrs. Creery's screams, and the mahout's imprecations! He had happened

to notice a banana-tree—he was extremely
partial to bananas !—and he made his way up
to it, tore off all bunches within his reach,
and devoured them with as much deliberation and satisfaction, as if there were not
seven furious, frightened, howling, screaming
human beings seated on his back. He flatly
refused to stir until he chose ! The policemen
were not within sight, and he seemed to be
tossing a half-penny in his own mind, as to
whether he would go for a ramble through
the jungle, or return to the path of duty,
which led to Mount Harriet and his afternoon rice. The afternoon rice had it, and
he accordingly strolled back, nearly tearing
his load off the howdah as he passed under
big branches—but that, he evidently considered, was entirely *their* affair—and then
climbed in a leisurely manner up the steep
bank he had recently descended, and resumed
the public road,—merely stopping now and
then, to snatch some tempting morsel, or to
turn round and round, in a very disagreeable
fashion. The fact was, he was not accustomed
to society, nor to carrying a load of pleasure-

seekers, and he did not like it. Dragging timber, and conveying stores was far more to his taste, and, besides this, Mrs. Creery's squeals, and her lively green umbrella, annoyed him excessively; he had taken a special dislike to her;—Chootie was not an amiable elephant, and would have thoroughly enjoyed tossing the lady with his trunk—and stamping on her subsequently. At last the party found themselves in front of the Mount Harriet bungalow, to their great relief and delight, and scrambled down a ladder, for, of course, their late conveyance would not condescend to kneel. Mrs. Creery, once safe on *terra firma*, was both bold and furious; and, standing on the steps, harangued the mahout in Hindustani on the enormity of the elephant's behaviour. She called him all the epithets she could immediately bring to mind, said she would complain to the General, and have him shipped to the Nicobars—that he was an ugly, unruly, untamed brute!

Naturally the elephant understood every word of this! (Hindustani is to them, as it were, their native language.) He calmly

waited till the irate lady had said her say
and furled (oh, foolish dame !) her umbrella;
and then he slowly turned his trunk in her
direction like a hose; there was a "whish,"
and instantly she, and her elegant costume,
were drenched from head to foot in dirty
water. What a spectacle she was! What a
scene ensued ! Vainly she fled ; the wetting
was an accomplished fact ; it had been very
sudden, and disastrously complete. Dr.
Malone actually lay down and rolled in the
grass, like the rude uncivilized Irish savage
that he was; Miss Caggett was absolutely
hysterical, and screamed like a peacock.
Helen and Mrs. Home, with difficulty re-
straining themselves, endeavoured to ame-
liorate the condition of the unhappy lady.
They escorted her inside the bungalow,
helped her to remove her gown, gloves, and
hat ; she was for once in her life actually
too angry to *speak*—she wept. Her dress
had to be despatched to the cook-house to
be washed and dried, and she, of course,
was in consequence prevented from taking
the head of the table, and had to have her

meal sent out to her in the retirement of the bedroom, where she discussed it *alone*. And the worst of it was, that she met with but little real sympathy. When she reappeared once more in public, she was met with wreathed smiles, and broad grins. Such is friendship! The company wandered about the hill after dinner, and Helen, thinking to checkmate James Quentin for once, offered her society to Dr. Parkes, who was only too pleased to accompany her—as long as she did not go too far, and there was no climbing. To punish Miss Denis for her want of taste, Apollo once more devoted himself to Lizzie,—being under the foolish impression, that, in so doing, he was searing Helen's very soul. It was soon tea-time; there was no moon, for a wonder; people had to depend on the stars and the fire-flies, and Mrs. Creery,—who had had a most disagreeable day,—gave the signal for an early departure. They all descended by a long, steep, winding pathway through the jungle, instead of by the more public road, as their boats were awaiting them at Hopetown pier; Mrs. Creery led the van, in a jampan carried

by four coolies—and, indeed, all the ladies
preferred this hum-drum mode of transport
to trusting themselves again to " Chootie,"
who was the bearer of some half-dozen
adventurous spirits, whom he took right
through the jungle, thereby reducing their
garments to rags, and covering their faces
with quite a pretty pattern of scratches ! Mr.
Quentin travelled per jampan, but Mr. Lisle
walked, and considered that he had much the
best of it ; so he had,—for he walked at Helen
Denis' right hand, and they both found this
by far the most delightful part of the day !
—whether this was due to the surrounding
influences, or to each other's society, I will
leave an open question. About a dozen
ticket-of-leave men accompanied the procession with flaring lights, as it wound down,
and down, the rugged pathway through the
forest, and gave the whole scene a fantastic
and picturesque appearance. It was a lovely
night, though moonless ; millions of silent
stars spangled the heavens, millions of fire-
flies twinkled in the jungle. Helen never
forgot that balmy tropical evening, with the

glow of torches illuminating the dark, lux-
uriant underwood, the scent of the flowers,
and the faint sound of the sea.

Mr. Lisle realized as he descended that
steep hill-path, that he was deeply in for it
at last, and in love with this Helen Denis,
helplessly in love—hopelessly in love—for he
might not speak, nor ever " tell his love;"
he could only play the part of confidant to
James Quentin, and, perchance, the thankless
rôle of best man !

Little did he guess, that the young lady
at his side, was not wholly indifferent to
him; that her blushes, when he appeared
with Jim, were to be put down to his own,
not to his companion's credit; that his mere
presence had the curious effect of abstracting
the interest from every one else, as far as
she was concerned—though, to be candid,
she never admitted this tell-tale fact to herself.
A gleam of the truth, a ray of rapture, came
to Gilbert Lisle by the flash of one of those
flaming torches,—was it imaginary? or was
it not? She smiled on him, as, he believed,
a girl only smiles on a man she cares for—

and yet Jim was absent.—Jim was yards behind, a leaden burthen to his lagging bearers.

A wild, ecstatic idea flashed through his mind, that she might—might not care for Quentin, after all! But this notion was speedily extinguished by his friend, who had noticed Lisle in attendance on Miss Denis on the way down the hill,—noticed that they stood a little apart on the pier before embarking, and neither " liked nor loved the thing he saw ! " Lisle the invulnerable was proof no longer. Lisle was a good-looking fellow, despite his shabby clothes and sunburnt skin. Yes, he had somewhat overlooked that fact. But Lisle was not a ladies' man, and he was a man of honour, and Mr. Quentin fully determined to give him to understand that he must not trespass on *his* preserves. Miss Denis belonged exclusively to him. And now let us privately examine Mr. Quentin's mind. Briefly stated, he did not " mean anything," in other words, he did not wish to marry her now—*that* fevered dream was past. He was not an atom in love with her either; she was too irresponsive, and,

in fact, too—as he expressed it to himself—
"stupid." Between ourselves, if any wandering damsel had appeared upon the scene,
he was ready to whistle Miss Denis down
the wind at once ! But damsels were rare at
Ross—and he still admired her greatly ;
he did not mean to " drop " her, till he went
away, and he intended to take precious good
care that no one should have it in their
power to say that *she* had dropped him—
much less, abandoned him for another. His
character as a lady-killer was at stake ; he
could not, and would not, lose what was as
precious to him as the very breath of his
nostrils.

He accordingly took an early opportunity of
giving Lisle what he called " a bit of a hint."

" I saw you making yourself very agreeable
to the fair Helen yesterday," he remarked
with affected *bonhomie.* " You mustn't make
yourself too agreeable, you know ! "

" Why not ? " demanded his companion
with exasperating composure.

" Why not ? My dear fellow, the idea of
your asking *me* such a question ! You know
very well why not."

" Am I to understand that she is engaged to you ? "

Mr. Quentin hated these direct questions, and why should Lisle look at him as if he were a witness that he was examining on his oath ?

" What is it to you ? " he returned evasively. " Come now, Lisle," leaning on his elbow, and smiling into the other's face with one of his most insinuating expressions.

" Answer my question first," roughly.

" Well, I will."

Word-fencing was easy to him, and he never thought it any harm to dissemble with a woman, and juggle his sentences so that one almost neutralized another; *they* were fair game, but a man was different. With men he could be frank enough—firstly, because he had more respect for his own sex; and secondly, because their eyes were not likely to be blinded by love, admiration, or vanity. Meanwhile, here was Lisle, an obstinate, down-right fellow, sternly waiting for his reply. An answer he must have, so he made a bold plunge, and said, with lowered eyelids and in a confidential voice,—

" What I tell you is strictly masonic, mind —but I know you are to be depended on. There is no actual engagement as yet between Helen and me,—but there is an understanding ! "

" I confess, the distinction is too subtle for me. Pray explain it ! "

" How can I go to her father whilst my money affairs are in such a confounded muddle ? Until I can do that, we cannot be what you call engaged. Do you see ? "

" I see. But there is one thing I fail to see —that Miss Denis treats you differently to any one else, or as if she were attached to you—in fact, latterly, it has struck me that she rather avoids you than otherwise !

This was a facer, but his companion was equal to the occasion. " That is easily explained," he replied. " She is the very shyest girl that you ever saw—in public."

Mr. Quentin thoroughly understood the art of inuendo, and the management of the various inflections of the human voice. He was a matchless amateur " star," and could " act " off, as well as on the stage.

After receiving this confidence, Mr. Lisle was silent; he leant back in his chair, and nearly bit his cigar in two. That last speech of Jim's had made him feel what the Americans call " *real* bad." A very long gap in the conversation ensued, and then he, as it were, roused himself once more,—

" Then she *is* engaged to you ! "

" No, not quite, not altogether—but our position is such, that no man of honour, knowing it, would take advantage of the situation,—would he ? "

" No—of course not."

And with this admission the subject dropped.

Mr. Quentin had succeeded brilliantly. He had assured Lisle that he was not engaged ; and yet he had impressed him with the fact that an engagement existed,—indeed, he had almost persuaded *himself*, that there was an understanding between him and Helen ! " Understanding " was a good, useful elastic word; it might mean an understanding to play tennis, to sit next each other at an afternoon tea, or to share the same umbrella !

"No, no, Mr. Gilbert Lisle," he said to himself exultantly, as he watched the other's gloomy face, "I'm not just going to let you cut me out—not if I *know* it. 'Paws off, Pompey.'"

CHAPTER III.

" STOLEN FROM THE SEA ! "

"Love, whose month is ever May,
Spied a blossom passing fair."
Much Ado About Nothing.

" Another fine, sunshiny day," is naturally of common recurrence in the East, and it was yet another magnificent afternoon at Ross— very bright, very warm, and very still. Underneath the long wooden pier vast shoals of little silver sardines were hurrying through the water, pursued by a greedy dolphin, and leaping now and then in a glittering shower into the air to escape his voracious jaws. Coal-black, stunted Andamanese, were here and there squatting on the rocks, patiently angling with the most primitive of tackle, and two or three policemen, in roomy blue tunics and portentous turbans, were gossiping together about rupees and rice. Some half-dozen soldiers, with open coat

and pipe in mouth, sat, with their legs
dangling over the pier, fishing. Further on,
with folded arms, and wistful eyes, a tall,
gaunt Bengalee stood, aloof and alone. He
was a zemindar from Oude, and had been in
the settlement since 1858, (an ominous date,)
now he was the holder of a ticket, was free
to open a shop in the bazaar, and make a
rapid fortune; free to accept a plot of the
most fertile ground on the face of the
globe, free to marry a convict woman, free
within the settlement, but there his liberty
ended. His body is imprisoned, but who
can chain the mind? His is far away beyond
those dim, blue islands, and the shining
"Kala Panee!" In imagination he now stands,
not upon Ross pier, but on wide-stretching
plains far north; his horizon is bounded by
magnificent forest trees, and topes of fragrant
mangoes; once more, he sees his native
village, and the familiar well, his plot of land,
his home; just as he saw it twenty years ago.
But too well does he remember every inmate
of those small, whitewashed hovels; their
faces are before him now—for, alas! what

has been left to *him* but memory ? Bitterly
has he expiated those few frenzied weeks, when
for a brief space, he and his neighbours felt
that they had broken the accursed yoke, and
trampled it beneath their feet—bitterer, ten
times, is it to know that he was sold and
betrayed by his own familiar friend !

At this maddening recollection, a kind of
convulsive spasm contracts his features, and
he mutters fiercely in his beard. He would
gladly—nay, gratefully—give all that remains
to him of life, just to have " Ram Sing " at
his mercy for one short moment, ay, but
one ! These are some of the thoughts that
flit through his mind, as he stands apart
with folded arms, and his dark, hawk-like
countenance, immovably bent on the sea,
deaf to the hoarse, loud laughter of Tommy
Atkins, who has had a good " take "—to the
screeching home-bound peacocks, and the
discordant yells of the Andamanese at play.

They have no tragic memories, this group
of young men coming down the pier in tennis
garb ; or, if they have, their faces much belie
them—Mr. Quentin, Captain Rodney, Mr.

Reid, and Dr. Malone (whose smooth, fair skin, and sandy hair, disavow his thirty summers).

"I told you so!" he exclaimed, as he hitched himself up on the edge of the pier. "They are all gone out, every man-Jack of them,—the Creerys, the Homes, Dr. Parkes, and Mr. Latimer, not to speak of our two young ladies. They have gone down to Chatham to take tea with Mrs. Grahame, and the island is a desert!"

"Fancy going three miles by water for a cup of hot water," said Mr. Quentin derisively; "but women will go *anywhere* for tea. Where are Jones and Lea?" he inquired.

"Where you ought to be, my boy; up decorating the mess for the dance this evening."

"Oh!" rather grandly, "I sent my butler over, and lots of flowers."

"If we were all to do that, I wonder 'what like it would be,' as they say in your native land, Reid?" remarked Dr. Malone. "And where is Green?"

"Out fishing with Lisle," replied Captain

Rodney. "And, ahem! talk of angels, here they come," as at this moment a sailing-boat suddenly shot round a point, and made for the pier.

"I've not seen Lisle for weeks!" remarked Dr. Malone; "not since the picnic on Mount Harriet. What has he been up to?"—to Mr. Quentin.

"Oh! he only enjoys society by fits and starts, and a little of it goes a long way with him."

"Hullo, you fellows!" hailed the doctor, leaning half his long body over the railings, "any luck?"

"Luck! I should just think so!" returned Lisle, standing up. "Two bottle-nosed sharks, a conger eel, a sword-fish, and any quantity of sea monsters, name and tribe unknown."

"Is that all?"

"No, not all. Green caught about a dozen crabs going out."

"Oh! now I say," expostulated Mr. Green, a fair young subaltern about six months from Sandhurst, "it was those beastly oars."

" There was an animal like a sea-cow, that nearly towed us over to Burmah," said Mr. Lisle, as he came up the steps, " and finally went off with all the tackle."

"The sea serpent, of course !" ejaculated Dr. Malone. "And, by the way, how is it that we have not seen you for a month of Sundays, eh ? Coming to the ball to-night ?"

" Ball ! what ball ? How can there be one without ladies ?"

"Nonsense, man alive ! what are you talking about ? Haven't we seventeen ?" putting his hat under his arm, and commencing to count on his fingers. "There is Mrs. King, Mrs. Grahame, Mrs. Manners—the widow from Viper—Mrs. Creery—"

"Mrs. Creery ! You may as well say Mrs. Caggett while you are about it."

"I may *not*. Mrs. Creery is a grand woman to dance, and you will see her and your humble servant taking the floor in style before you are many hours older ! If all the ladies put in an appearance, and do their duty, we shall have an A 1 dance ! Of course you are coming ?"

"No," put in Mr. Quentin, rather quickly.
"How could you ask him? Does he look
like a dancing man? Here are the fish
coming up? What whoppers!" turning
towards the steps.

"And here comes something else!" exclaimed the doctor, pointing to a white sail
approaching the island. "It's easy to see
what *you* have come down for, my boy!"
to Apollo, who smiled significantly, and
accepted the soft impeachment without
demur.

"Quentin is a lucky fellow, isn't he?"
said Mr. Green, addressing himself to Mr.
Lisle, with all the enthusiasm of ignorance.
"He has had it all his own way from the
first; none of us were in it! And although
our circle of ladies *is* small, I'll venture to
say we could show a beauty against Madras
or Rangoon; yes, and I'll throw in Calcutta too! I'll back 'La Belle Hélène'
against anything they like to enter, for pace,
shape, and looks!"

Here Mr. Lisle turned upon his heel, and
walked away.

"What's up? What's the matter, eh?" demanded the youth of Mr. Quentin, who was now gazing abstractedly at the approaching boat, with a cigarette between his teeth.

"Oh, he did not approve of your conversation; he does not think that ladies should be talked about, and all that sort of rubbish."

"Pooh, why not?—and was I not praising her up to the skies? What more could I have said? And I'm sure, if you don't mind, *he* need not!"

"No, but he did," remarked Dr. Malone. "He looked capable just now of tossing you out as a sort of light supper to the sharks, my little C. Green!"

"And a very light meal it would be," said Mr. Green with a broad grin. "Nothing but clothes and bones. Here comes Miss Caggett and a whole lot of people, and won't she just walk into *us* for not decorating the mess!"

At this instant Miss Caggett and some half-dozen satellites appeared in view, and behind her, walking with Dr. Parkes, came a lady we have never seen before, Mrs. Durand, who had only that morning returned to the settlement.

"Well," cried the sprightly Lizzie, surveying the guilty group with great dignity, "I call this *pretty* behaviour! What a lazy, selfish, good-for-nothing set!" beginning piano, and ending crescendo.

. Dr. Malone nodded his head like a mandarin at each of these adjectives, and declared,—

"So they *are*, Miss Caggett, so they are. I quite agree with you."

The young lady merely darted a scornful glance in his direction, and proceeded,—

"Mr. Quentin, well, I've given you up long ago! Mr. Green, I cannot say much to *you*, when grown-up people set you such an example" (a back-handed slap at Mr. C. Green's tender years). "Mr. Lisle, you here? and pray what have you got to say for yourself? What is your excuse?"

"My excuse," coming forward and doffing his hat, "is, that I have no more idea of decorating a room than one of the settlement elephants—in fact, my genius is of a destructive, rather than a constructive order. But I am always prepared to appreciate other people's handiwork."

"Well, you *are* cool," staring at him for a second in scornful silence.

"Now, Dr. Malone," pointing at him with her parasol, "let us hear what you have got to say for yourself."

Dr. Malone rested his chin on the top of his tennis-bat, and calmly contemplated his fair questioner, in a somewhat dreamy fashion, and then was understood to say,—

"That as long as Miss Caggett was in a ball-room, any other decoration was quite superfluous!"

To which Miss Caggett responded by rapping him on the knuckles with the handle of her sunshade, and saying,—

"Blarney!"

Meanwhile Mrs. Durand had joined the group, and now received a very warm welcome. It was easy to see that she was a popular person at Port Blair. She was upwards of thirty, with a full, but very erect figure, smiling, dark eyes, good features, and white teeth, the upper row of which she showed very much as she talked. She wore a hat with a dark blue veil, a pretty cambric

dress, and carried a red parasol over her arm
(a grand landmark, that same parasol, for
Mrs. Creery).

"Great events never happen alone!" quoth
Dr. Malone, bowing over his bat. "Here, in
one day, we have the mail in, the full moon,
the ball, and Mrs. Durand! It is quite
needless to inquire after Mrs. Durand's
health ?"

Mr. Quentin moved forward to accost the
lady, his large person having hitherto entirely concealed his friend, and as he moved,
Mrs. Durand's eyes fell upon Gilbert Lisle.
She opened them very wide, shut them, and
opened them once more, and said in a slow,
staccato voice,—

"I believe I am not dreaming, and that I
see Mr. Lisle. Mr. Lisle," holding out a
plump and eager hand, "what on *earth*
brings *you* here ?"

Precisely what every one wanted to
know.

Mrs. Durand had a habit of laying great
stress on some of her words, and she uttered
the word earth with extraordinary emphasis.

Her acquaintance, upon whom all eyes were now riveted, smiled, shook hands, muttered incoherently, and contrived, by some skilful manœuvre, to draw the lady from the centre of the crowd.

"I never was so amazed in my life!" she reiterated. "What put it into your head to come here, of all places?"

"Oh, I wanted to see something out of the common, and to enlarge my ideas."

"Indeed, I did not know that they required extension! One could understand our being here—we are sent, like the convicts; but outsiders—and, of all people, you!"

"There is first-class fishing to be had, and boating, and all that sort of thing; and the scenery is perfect," he answered.

"Granted—and pray how long have you been at Port Blair?"

"I came in July," he replied, rather apologetically.

"July!" she echoed, "and this is November!—*five* months! And may I ask what is the attraction, besides sailing and sharks?"

"The unconventional life, the temporary

escape from politics and post-cards, express
trains, telegrams, and the bores of one's
acquaintance."

" Well, every one to their taste, of course !
You like Port Blair, give *me* Park Lane. As
to politics, we have our politics here. Have
you not discovered that we are an absolute
monarchy ?"

" Yes," smiling ; " but, alas ! I am not in
favour at court."

" No ? neither am I. I'm in the Oppo-
sition. I'm one of the Reds," laughing, and
displaying all her teeth. " Here are all these
people coming back, and I must go ; I have
a great deal to do at home. Remember, that
I shall expect to see you very often—*sans
ceremonie*. Oh, I suppose that tall girl is
Miss Denis ? Charlie says she is uncommonly
pretty, and not spoiled *yet*. By the way,"
pausing, and looking at him significantly,
" I wonder if you have been losing your
heart, as well as enlarging your ideas ?"

" Do I ever lose my heart ?" he asked.
" Am I an inflammable person ?"

" No, indeed—quite the reverse; warranted

not to ignite, I should say," shaking her head.
"And now I really must be going, or Mrs.
Creery will catch me, and cross-examine me.
Of course, we shall meet this evening?"
Mr. Lisle walked with her to the end of the
pier, bending towards her, and apparently
speaking with unusual earnestness, as Miss
Caggett remarked. At the gate, he and the
lady parted, he taking off his hat, she waving
her hand towards him twice, as if to enforce
some special injunction.

The gig was now alongside the steps, and
its late passengers had ascended to the pier.
Miss Denis was the last to leave the boat,
and was at once surrounded by Mr. Quentin,
Dr. Malone, Captain Rodney, and Mr. Green,
a faithless quartette, who all quitted Miss
Caggett in a body.

"Well, Miss Denis," said Mr. Green, "I
am glad to see that you have not forgotten
the button-hole I asked you to bring me,"
pointing to a flower in the front of her dress.

"Oh, this!" taking it out and twirling it
carelessly in her fingers. "I certainly did
not gather it for your adornment, but still,

if you like," half tendering it ; but becoming
conscious of Mr. Quentin's greedy, out-
stretched hand, she paused.

"You surely would not ? " he began pathe-
tically.

"No, I would *not*, certainly not. I will
give it to the sea," and suiting the action to
the word, she tossed it over the railings into
the water.

"Oh, Miss Denis," exclaimed Mr. Green
with a groan, "how could you trifle with
my feelings in such a manner ? How could
you raise me to a pinnacle of happiness, and
cast me down to the depths of despair? Have
you no conscience ? "

"It would have been a precedent," she
answered gaily. "I know you only too
well—you would have demanded a bouquet
every time I returned to the island."

Here, for the first time, her eyes fell upon
Mr. Lisle, who had now joined the outer
circle—Mr. Lisle, whom she had not seen
for six weeks. She coloured with astonishment, and accorded him rather a stiff little
bow. He did not come forward, but con-

tented himself with merely raising his hat, and remaining in the background.

Helen had once rather timidly asked after him, from Mr. Quentin (it seemed so strange, that he had never been over to Ross, since the day of the picnic, when they had made that never-to-be-forgotten expedition down the mountain, escorted by torches and fire-flies).

To Miss Denis's somewhat faltering question, Mr. Quentin had brusquely replied " that Lisle had on one of his sulky fits, and the chances were, he would not come over to Ross again—he was an odd, unsociable, surly sort of beggar ! "

Apparently he had now recovered from the sulks ; for there he stood, looking as sunburnt, as shabby, and as self-possessed as ever !

" We had a pleasant sail," remarked Mrs. Creery, " but I could not go in at Chatham on account of Nip ! Mrs. Grahame makes such a fuss about that hideous puppy of hers— and, after all, it's only Nip's play ! Of course, I could not leave the poor darling in the boat by himself, so we had our tea sent out to us, and were very happy all the same," hugging him as she spoke with sudden rapture.

But Nip (whose *play* was death to other dogs) stiffened his spine, and threw back his head; he evidently considered public endearments inconsistent with personal dignity. He weighed fully twenty-four pounds, and why Mrs. Creery carried an animal who had the excellent use of his four legs, was best known to herself.

As she proceeded up the pier, with his head hanging over her shoulder, he surveyed Dr. Malone and Lisle, who walked behind him, with unconcealed contempt.

"What a fool she makes of herself about that beast!" muttered the former. "He despises *us* for not being carried too. I see it in his eye! Brute! I'd like to vivisect him."

"Only imagine!" exclaimed Miss Caggett suddenly, "Miss Denis has never been to a dance in her life!—and," giggling affectedly, "never danced with any but *girls*."

"And remember," said Jim Quentin, impressively turning and speaking to Helen in a tender undertone (for the benefit of his friend), "that you have given *me* the promise of the first waltz."

The party had now reached a little square, from whence their various paths diverged.

"You wait for me on the pier like a good fellow," he said to his companion. "I am just going to walk home with Miss Denis."

Every one now departed in different directions, excepting Mrs. Creery, who remained behind at the cross-roads, for a moment, and waving her green umbrella, called after them authoritatively,—

"Now mind that none of you are *late* this evening!—you men especially!"

.   .   .   .   .

Mr. Lisle went slowly back to the pier; it was almost deserted now. Tommy Atkins had adjourned to his well-earned supper, the jailor to his rice, the Andamanese to unknown horrors. The zemindar is alone—alone he stands, and sees what is to him another wasted sun sink into the sea like a ball of crimson fire! Apparently he is unconscious of a figure, who comes and leans over the

railings, with his eyes fixed abstractedly on
the sea, till with a sudden flash they become
riveted on something, scarcely deserving
such eager inspection—merely a floating
flower! As Gilbert Lisle gazed, he was the
prey of sore temptation. Surely, he argued
with himself, there would be no harm in
picking up a castaway lily, even Quentin
would hardly grudge him that, and *he* might
as well have it as the sea! Then he turned
half away, as if thrusting the impulse from
him (the convict now noticed him for the
first time); but the flower was potent, and
drew him back; he leant his arms on the
railings, and stared at it steadily. The
zemindar watched him narrowly out of his
long, black eyes. The Sahib was debating
some important question in his own mind! he
looked at his watch, he glanced nervously up
and down the pier, apparently his companion
was as naught. Then he hurried to the foot
of the steps and unmoored a punt, and rowed
out several lengths, in quest of *what?* A
white flower that the tall English girl had
thrown away.

The native followed his quest with scornful interest. He has it now ;—no, it has evaded him, and still floats on. Ah, he has reached it this time, he has lifted it out of the water, as reverently as if it were one of the sacred hairs of Buddha! He has dried it; he has concealed it in his coat!

Bah! the Feringhee is a fool!

CHAPTER IV.

THE BALL.

"There was a sound of revelry by night."

NIGHT had fallen, and the full moon to which Dr. Malone had alluded was sailing overhead, and flooding Ross with a light that was almost fierce in its intensity; the island seemed to be set in a silver sea, over which various heavily laden boats were rowing from the mainland, conveying company to the ball! Jampans bearing ladies were to be seen going up towards the mess-house in single file, the guests kept pouring in, and, despite the paucity of the fair sex, made a goodly show! We notice Mrs. Creery (as who would not?) in a crimson satin, with low body, short sleeves, and a black velvet coronet on her head. Helen Denis in white muslin, with natural flowers; she had been forbidden by the former lady to even so much as *think* of her white silk, but had, nevertheless, cast many yearnings in that direction. All the same, she looks as well as her best friends could wish, and a certain nervousness and anticipation gives unwonted brilliancy to her colour (indeed, Miss Caggett has already whispered " paint! "), and unusual brightness to her eyes.

The world seems a very good place to her this evening. She is little more than eighteen, and it is her first dance; if she has an *arrière pensée*, it has to do with Mr. Lisle, who after being so—well, shall we say " interesting ? " and behaving so heroically, has calmly subsided into his normal state, viz. obscurity. What is the reason of it? Why will he not even speak to her ? Little does she guess at the real motive of his absence. As little as that, during his long daily excursions by land and sea, a face, *hers*, forms a constant background to all his thoughts—try and forget it as he will.

The mess-room looked like a fairy bower, with festoons of trailing creepers and orchids twined along the walls, with big palms and ferns, in lavish profusion, in every available nook. It was lit up by dozens of wall-lamps, the floor was as smooth as glass, and all the most comfortable chairs in Ross were disposed about the ante-room, and verandahs.

The five-and-forty men were struggling into their gloves, and hanging round the door, as is their usual behaviour, preliminary to a dance; and the seventeen ladies were scattered about, as though resolved to make as much show as possible. Mrs. Creery occupied a conspicuous position; she stood exactly in the middle of the ball-room, holding converse with the General, who bowed his head acquiescently from time to time, but was never so mad as to try and get in a word edgeways. Nip was seated on a sofa, alert and wide awake, plainly looking upon the whole affair as tomfoolery and nonsense; but he had been to previous entertainments, and knew that there was such a thing as *supper!*

Near the door, stood Miss Caggett, the centre of a noisy circle, dangling her programme, and almost drowning the bass and tenor voices by which she was encompassed, with her shrill treble, and shrieks of discordant laughter at Dr. Malone's muttered witticisms. Her dress was pink tarletan, made with very full skirts, and it fitted her neat little figure to perfection. Altogether, Miss Caggett was looking her best, and was serenely confident of herself, and severely critical of others.

Every one had now arrived, save Mr. Quentin, but he thoroughly understood the importance of a tardy and solitary *entrée*. At last his tall figure loomed in the doorway, and he lounged in, with an air of supreme nonchalance, just as the preliminary bars of the opening Lancers were being played.

He was not alone, to every one's amazement he was supplemented by Mr. Lisle—Mr. Lisle in evening dress! There had been grave doubts as to his possessing that garb ; and his absence from one or two dinners had been leniently attributed to this deficiency in his wardrobe ! People who looked once at James Quentin, looked twice at Gilbert

Lisle; they could hardly credit the evidence of their senses. Mr. Lisle in unimpeachable clothes, with a matchless tie, a wide expanse of shirt-front, and skin-fitting gloves, was a totally different person to the individual they were accustomed to see, in a rusty old coat, a flannel shirt, and disreputable wide-awake! How much depends on a man's tailor! Here was the loafer, transformed into a handsome (if rather bronzed,) distinguished-looking gentleman. He received the fire of many eyes with the utmost equanimity, as he leant lazily against the wall, like his neighbours. Miss Caggett, having breathed the words " Borrowed plumes," and giggled at her own wit, presently beckoned him to approach, and said pertly,—

" This is, indeed, an unexpected pleasure. I thought you said you were not coming, Mr. Lisle ? "

" Did I ? " pausing before her. " Very likely; but, unfortunately, I am the victim of constitutional vacillation."

" In plain English, you often change your mind ? "

E 2

"*Never* about Miss Caggett," bowing deeply, and presently retiring to the doorway.

Lookers-on chuckled, and considered that "Lizzie," as they called her among themselves, had got the worst of *that!* Mrs. Creery, who had been gazing at this late arrival with haughty amazement, now no longer able to restrain herself, advanced upon him, as if marching to slow music, and said,—

"I've just had a letter about *you*, Mr. Lisle."

Mr. Lisle coloured—that is to say, his tan became of a still deeper shade of brown, and his dark eyes, as they met hers, had an anxious, uneasy expression.

"Oh, yes!" triumphantly, "I know *all* about you, and who you are, and I shall certainly make it my business to inform every one, and—"

"Do not for goodness' sake, Mrs. Creery!" he interrupted eagerly. "Do me the greatest of favours, and keep what you know to yourself."

Mrs. Creery reared back her diademed

head, like a cobra about to strike, and was on the point of making some withering reply, when the General accosted her with his elbow crooked in her direction, and said, " I believe this is our dance," and thus with a nod to her companion, implying that she had by no means done with him, she was led away to open the ball.

Meanwhile Helen had overheard Mrs. Grahame whisper across her to Mrs. Home,—

" What do you think ? When Mrs. Creery came back from us, she found her letters at home, and she has heard something *dreadful* about Mr. Lisle ! "

Helen was conscious of a thrill of dismay as she listened. She was so perplexed, and so preoccupied, that she scarcely knew what she was saying, when Mr. Quentin came and led her away to dance. During the Lancers she was visibly *distrait*, and her attention was wandering from the figures and her partner, but she was soon brought to her senses by Mr. Quentin saying rather abruptly,—

" I've just heard a most awful piece of

news!"—her heart bounded. "Only fancy their sending *me* to the Nicobars!"

Helen breathed more freely as she stammered out,—

"The Nicobars?"

"Yes, the order came this evening by the *Scotia*—sharp work—and I sail in her for Camorta to-morrow at cock-crow."

"And must you go, really?"

"Yes, of course I must. Isn't it hard lines? Some bother about the new barracks. The Nicobars are a ghastly hole, a poisonous place. I shall be away two months—that is, if I ever come *back*," he added in a lachrymose voice.

"And what about Mr. Lisle?"

"Oh, he is such a beggar for seeing new regions—he is coming too."

"I'm sorry you are going to the Nicobars, they have such a bad name for fever and malaria."

"I believe you! I hear the malaria there rises like pea-soup!"

"Mr. Lisle is foolish to go; you should not let him."

"Oh! he may as well be there as here! He is as hard as nails, and it would be deadly for me without a companion. He promised to come, and I shan't let him off, though I must confess, what he *says*, he sticks to."

Miss Denis thought Mr. Quentin's arrangements savoured of abominable selfishness, and between this news, and the sword of Damocles that was swinging over Mr. Lisle's head, her brain was busy. Dancing went on merrily, but she did not enjoy herself nearly as much as she anticipated. After all, this apple of delight, her first ball, had turned to dust and ashes in her mouth. And why?

Mr. Lisle leant against a doorway, and looked on very gravely: doubtless he knew the fate that was in store for him. He remained at his post for the best part of an hour, and had any one taken the trouble to watch him, they would have noticed that his eyes followed Helen and Jim Quentin more closely than any other couple. As they stopped beside him once, she said,—

"I did not know that you were coming to-night, Mr. Lisle."

"Neither did I, till quite late in the afternoon. I suppose there is not the slightest use in my asking for a dance?"

Now if the young lady had been an experienced campaigner, and had wished to dance with the gentleman (which she did), she would have artlessly replied,—

"Oh, yes! I think I can give you number so and so," mentally throwing over some less popular partner; but Helen looked straight into his face with grave, truthful eyes, displayed a crowded programme, and shook her head.

Jim Quentin, who was evidently impatient at this delay, placed his arm round his partner's waist, and danced her away to the melting strains of the old "Kate Kearney" waltz.

None gave themselves more thoroughly up to the pleasures of the moment, or with more "abandon" than Dr. Malone and Mrs. Creery. They floated round and round, and to and fro, with cork-like buoyancy, for

Mrs. Creery, though elderly and stout, was
light of foot, and a capital dancer; and her
partner whirled her hither and thither like a
big red feather! Every one danced, and the
seventeen revolving couples made quite a
respectable appearance in the narrow room.
And what a sight to behold the twenty-eight
partnerless men, languishing in doorways, and
clamouring for halves and quarters of dances!
Men who, from the wicked perversity of their
nature, were they as one man to ten girls,
would certainly decline to dance at *all!* Mr.
Lisle had abandoned his station at last, and
waltzed repeatedly with Mrs. Durand; they
seemed to know each other intimately, and
were by far the best waltzers in the room.
There was a finish and ease about their per-
formance that spoke of Balls in the Great
Babylon, and though others might pause
for breath, and pant, and puff, these two,
like the brook, seemed to " go on for ever !"

They also put a very liberal interpretation
upon the term " sitting out!" They walked
up the hill in the moonlight, and surveyed
the view—undoubtedly other dancers did the

same—but not *always* with the same companion ; to be brief, people were beginning to
talk of the "marked" attention that Mr.
Lisle was paying Mrs. Durand—attentions not
lost on Helen, who noticed them, as it were,
against her will, and tried to keep down a
storm of angry thoughts in her heart by
asking herself, as she paced the verandah
with Dr. Parkes, and dropped hap-hazard
sentences, "Was it possible that she was
jealous, bitterly jealous, because Mr. Lisle
spoke to another woman ?—Mr. Lisle, who
avoided her ; Mr. Lisle, who had a history ;
Mr. Lisle, who was going away ?"

She held her head rather higher than usual,
pressed her lips very firmly together, and told
herself, "No, she had not *yet* fallen quite so
low. Mr. Lisle and his friends were nothing
to her."

. . . . .

Supper was served early. Mrs. Creery
was the hostess, and we know that she had
"Nip" in her mind, when she suggested that
at twelve o'clock they should adjourn for re-
freshment, and sailed in at the head of the

procession on the General's arm. "Nip," who had been the first to enter the supper-room, sat close to his doating mistress, devouring tid-bits of cold roast peacock, and *pâté de foie gras*, with evident relish; *this* was a part of the entertainment that he could comprehend. His mistress was also pleased with the refection, and condescended to pass a handsome encomium upon the mess-cook, and priced several of the dishes set before her (with an eye to future entertainments of her own). She was in capital spirits, and imparted to Dr. Malone, who sat upon her left, that she had never seen a better ball in Ross in all her experience; also, amongst many other remarks, that Miss Caggett's dress was like a dancer's."

"But is not that as it ought to be?" he inquired, with assumed innocence.

"I mean a columbine!" she replied sternly; "and her face is an inch deep in powder—she is a *show!* As to Helen Denis—"

"Yes, Mrs. Creery. As to Miss Denis?"

"I'm greatly disappointed in her. She is no candle-light beauty, after all."

"Ah, well, maybe she will come to *that* by-and-by. So long as she can stand the daylight, there is hope for her—eh?"

Mrs. Creery told Dr. Malone that "she believed he was in love with the girl, or he would not talk such nonsense!" and finally wound up the conversation by darkly insinuating something terrible about Mr. Lisle, adding that he had craved for her forbearance, and implored her to hold her tongue!

"But I won't," she concluded, rising as she spoke, and dusting the crumbs off her ample lap. "It is my *duty* to expose him! We don't want any wolves in sheep's clothing prowling about the settlement," and with a nod weighty with warning, she moved away in the direction of the ball-room.

Miss Caggett had torn her dress badly— her columbine skirts—and Helen was not sorry to be called aside to render assistance. She was unutterably weary of Mr. Quentin and his monotonous compliments. His manner of protecting, and appropriating her, as if she belonged to him, and they had

some secret bond of union, was simply
maddening! As she tacked up Lizzie's
rents, in a corner of the ante-room, Lizzie
said suddenly,—

"I suppose you have heard all the fuss
about Mr. Lisle? Mrs. Creery is bubbling
over with the news. Don't pretend *I* told
you, but she has heard all about him at last;
very *much* at last," giggling.

"Yes?" interrogated her companion.

"He was in the army—I always suspected
that; he looked as if he had been drilled.
He was turned out, cashiered for something
disgraceful about racing; and as to his flirtations, we can imagine *them*, from the way
he is behaving himself to-night! He has
danced every dance with Mrs. Durand, though
I will say this, she asked him; and, of course,
it was because *she* came back, that he
changed his mind about the ball."

"Now your dress will do, I think," said
Helen, rising from her knees with rather a
choking sensation in her throat.

"Oh, thanks awfully, you dear girl!"
pirouetting as she spoke. "I'll do as much

for you another time; there's a dance beginning, and I must go!" and she hurried off.

In the doorway Helen came face to face with Mr. Lisle, who was apparently searching for some one,—for her!

She held up her chin, and, with one cool glance, was about to pass by, when he said, rather eagerly,—

"Miss Denis, I was looking for you. Malone has been sent for to barracks, and he said that I might ask you to give me his dance—the next—the last."

Helen fully intended to decline the pleasure, but something in Mr. Lisle's face compelled her to say " *Yes*," and without a word more, she placed her hand upon his arm ; they walked into the ball-room, and immediately commenced to waltz; this waltz was " Soldate Lieder." Her present partner was very superior to Jim Quentin, and she found that she could go on much longer with him without stopping, keeping up one even, delightful pace; but at last she was obliged to lean against the wall—completely out of breath. Her eyes, as she did so, followed Mrs. Durand enviously, and she exclaimed,—

"I wish I could dance like her." Now, had she breathed this aspiration to Mr. Quentin or Dr. Malone, they would have assured her that her dancing was already perfection, but Mr. Lisle frankly replied,—

"Oh, all you want is practice; you must remember that she has been at it for years. We used to dance together at children's parties,—I won't say *how* long ago."

"I know I dance badly," said Helen, colouring; "but the reason of that is that, although I danced a great deal at school, it was always as gentleman, because I was tall."

"Ah! I see," and he laughed. "Now I understand why you were so bent on steering me about just now. Well, you are not likely to dance gentleman again, I fancy. There," regretfully, "it's over; shall we go outside?"

Helen nodded her head, and accordingly they went down the steps arm in arm. She meant to seize this opportunity of giving him a hint of the mine on which he was standing, —one word of warning with regard to Mrs. Creery.   She had accepted his friendship, and surely this would be the act of a friend.

Mr. Quentin—sitting in the dusky shades of a secluded corner, whispering to Lizzie Caggett—saw the pair descending from the ball-room, pass down the steps, and out into the moonlight, and looked after them with an expression of annoyance that was quite a revelation to his sprightly companion.

CHAPTER V.

" BUT WHAT WILL PAPA SAY ?"

" Joy so seldom weaves a chain
Like this to-night, that, oh ! 'tis pain
To break its links so soon."

Moore.

HELEN and her partner ascended the steep
gravel pathway, lined with palms, gold
mohur, and orange-trees, and turning a sharp
corner, came suddenly upon a full view of the
sea, with the moon on her bosom. It was a
soft, still, tropical night; not a sound broke
the silence, save a distant murmur of human
voices, or the dip of an oar in the water.

That moon overhead seldom looked down
upon fairer scene, or a more well-favoured
couple, than the pair who were now leaning
over the rustic railings, and gazing at the
prospect beneath them—or rather, the man
was looking at the girl, and the girl was

looking at the sea. Doubtless moonshine idealizes the human form, just as it casts a glamour over the landscape ; but at the present moment Helen appears almost as beautiful as her world-renowned namesake. Her lovely eyes have a fathomless, far-away expression, her pure, clear-cut profile is thrown into admirable relief by the glossy dark leaves of a neighbouring orange-tree. In her simple muslin dress, with its soft lace ruffles, and a row of pearls round her throat, she seemed the very type of a modest English maiden (no painted columbine this !), and, perhaps, a little out of place amid her Eastern surroundings. She continued to gaze straight before her, with her hands crossed on the top of the railing, and her eyes fixed on the sea. As she gazed, a boat shot out of the dim shadows, and across the white moonlit track, then passed into obscurity again.

" Thinking as usual, Miss Denis ? " said her companion.

" Yes," she answered rather reluctantly, " thinking of something that I must say to *you*, and wondering how I am to say it."

"Is it much worse than last time?" he inquired with a smile (but there was an inflection of eagerness in his voice).

"Oh! quite different."

"Ah, she is going to announce that she is engaged to Quentin," he said to himself with a sharp twinge.

"Do you find it so very hard to tell me?" he inquired in a studiously indifferent tone.

"Yes, very hard; but I must. I owe you much, Mr. Lisle—and—I am your—friend—I wish to warn you." Suddenly sinking her voice to a whisper, she added,—"Mrs. Creery has had a letter about you!"

"Containing any startling revelations, any bad news?"

"Yes," she returned faintly. "Bad news. Oh, Mr. Lisle,—I am so sorry!"

"Is the news too terrible to be repeated?" he asked with marked deliberation.

Helen fidgeted with her fan, picked a bit of bark off the railing in front of her, and, after a long silence, and without raising her eyes, she said,—

"Must I tell you?"

F 2

" If you please," rather stiffly.

" She—she—hears that you have been in the army."

" Yes, so I was.—I was not aware that it was criminal to hold her Majesty's commission ; but, of course, Mrs. Creery knows best."

" She says you were—were obliged to— to leave disgraced," continued his companion in a rapid, broken whisper.

" Cashiered, you mean, of course ! "

" Yes," glancing at him nervously.   To her amazement, he was smiling.

" Do you believe this, Miss Denis ? " he asked, raising himself suddenly from a leaning posture and looking at her steadily.

" No," she faltered.   " I think not.   No," more audibly, " I do not," blushing deeply as she spoke.

" Why ? " he asked rather anxiously.

" I cannot give you any reason," she stammered, somewhat abashed by the steadfastness of his gaze, "except a woman's reason, that it is so—"

" I am sincerely grateful to you, Miss Denis ; your confidence is not misplaced.—I

am *not* the man in question. Mrs. Creery
has got hold of the wrong end of the stick
for once. I know of whom she is thinking,"
his face darkened as he spoke, " a namesake
and, I am ashamed to say, a relation of mine.
It is extremely good-natured of the old lady, to
make me the subject of her correspondence."
Then in quite another tone he said, " I suppose
you have heard of our start to-morrow ?"

"Yes," she replied, scarcely above a whisper.

" I'm a regular bird of passage, and ought
to have been away weeks ago ; and you yourself will probably be on the wing before long."
(He was thinking of her marriage with Jim
Quentin, but how could she know that ?)

" Oh, not for a year at any rate ! Papa
does not expect that we shall be moved before
then," she answered quite composedly. " I
am sorry you are going to the Nicobars—I
mean, you and Mr. Quentin," hastily correct-
ing herself. " It's a horribly unhealthy place
—soldiers and convicts die there by dozens
from—fever," her lip quivered a little as she
spoke.

" Not quite so bad as you think," returned

her companion, moving his elbow an inch closer to her. "I'm an old traveller, you know,—and I will look after him for you."

"Look after who?" she asked in amazement.

"Why, Quentin, to be sure. I know all about it. I," lowering his voice, "am in the *secret*."

"Mr. Lisle, will you kindly tell me at once what you mean?"

"Certainly, Miss Denis. I mean that Quentin is the happiest of men."

"I am extremely pleased to hear it, but why?" she interrogated firmly.

"What is the use of fencing with me in this way?" he exclaimed with a gesture of impatience. "You may trust me.—I know all about it. Quentin has told me himself, that he is engaged to you."

"Engaged to *me*!" she echoed with glowing eyes. "Mr. Lisle, you are joking."

"Do I look as if I was joking?" he demanded rather bitterly.

"It is not the case. It is the first that I have heard of it," exclaimed the young lady

in a voice trembling with agitation and indignation. " How dared he say so ? "

Mr. Lisle felt bewildered; a rapturous
possibility made his brain reel. Yet who
was he to believe ? Quentin had been very
positive; he had never known him to utter
a deliberate lie. And here, on the other
hand, stood this girl, saying " No;" and if
ever the truth was traced upon proud, indignant lips, it was written on hers.

" Do you believe me, Mr. Lisle ? " she
asked impatiently.

For fully a moment he did not speak;
and was it the moonlight, or some sudden
emotion, that made him look so white ?

" I do believe you, of course," he answered
in a low voice. " And now," he continued
in the same low tone, urged to speak by an
irresistible impulse, " perhaps you can guess
why I have stayed away? How, from a sense of
mistaken loyalty, my lips have been locked ? "

Her eyes, which up to this, had been fixed
intently on his, now sank. Suddenly a suspicion of the truth now dawned upon her
mind, and she turned aside her face.

" Miss Denis," he said, " I see you have guessed my secret—I love you."

These three magic words, were almost inaudible ; barely louder than the orange leaves which whispered in the scented air. Nevertheless a busy little zephyr caught them up, carried them away, and murmured them to the sleepy flowers and the drowsy waves, that washed the invulnerable rocks beneath them.

Helen made no reply. This was the first love tale to which she had ever listened, and those three syllables stirred every fibre of her heart.

" Do you remember, that time on the wreck," he continued, " when you told me that I was leading a lazy useless life, and that I ought to go back to the outer world ? You little guessed that it was you, yourself, who were keeping me a prisoner here ! "

Still the young lady said nothing, but kept her face steadily turned towards the sea.

He waited a moment, as if expecting some reply, but none came. At last he said, in quite a different tone,—

" I see how it is.—I have been a presumptuous idiot ! And, after all, I had no

right to expect that you would care a straw
about me. I am years older than you are;
I am—"

"Mr. Lisle," she interrupted, turning
towards him at last, and speaking with
apparent effort, "you are quite wrong.—I
—I," she stopped, and a little half-frightened
smile played round her mouth, as she added,
almost under her breath, "But what will
papa say?"

"Then *you* mean to say 'Yes'!" he
exclaimed, coming nearer to her, and grasping
the railing firmly in his hand, to conceal how
it shook.

Again she made no reply, but this time
Mr. Lisle undoubtedly took silence for
consent!

Mrs. Creery and Dr. Parkes were standing
on the very summit of the hill, overlooking
everything and everybody, and the former,
had not failed to notice a couple at some distance below them, leaning over the rails, and
contemplating the sea: a tall girl in white,
Helen Denis, of course; and who was the
man? It looked like Captain Durand.

There, Captain Durand had just bent over her, and kissed her hand! Pretty doings, certainly, for a married man.

"There!" she exclaimed, suddenly nudging Dr. Parkes, "did you see *that?*"

"See what, my dear madam?"

"That man down there with Helen Denis. I believe it's Captain Durand; he has just kissed her hand. Oh! WAIT till I see his wife!"

"Pooh!" returned her companion contemptuously, "the moonlight must have deceived you, it was his own hand; he was stroking his moustache."

"Oh, well, I'm not so sure of that!—but I suppose I must take your word for it, doctor."

Meanwhile, to return to Mr. Lisle, who *had* kissed Helen's hand. (Mrs. Creery's eyes seldom deceived her.) "Won't you say something to me, Helen?" he pleaded anxiously.

"Yes," turning round and drawing her fingers away, "I will.—I say—don't go to the Nicobars."

"But I must, I have promised Quentin

and Hall, and I cannot break my word. I
would gladly give half I possess, to get out
of it; but I little guessed this afternoon, when
Quentin .asked me to go, and I said 'Yes,'
that I would so soon have such very strong
reasons for saying ' *No.*' "

" I wish they would let you off; I have a
presentiment about the Nicobars."

" Presentiment of what ?"

" I cannot say, but of something bad. Do
you believe in presentiments ? " looking at
him wistfully.

" No, and yet I should not say so ! That
night of the storm, when you ran down the
pier steps and called me back, your voice
and your face haunted me afterwards for
days. I had a kind of conviction that I had
met my fate, and so *I had,* you see ! By the
way, I wonder why you like me, Helen ? or
what you see in me ? "

The young lady smiled, but said nothing.

" All the world can understand my caring
for you, but I am, in one way, an utter
stranger; you could not answer a single
question about me, if you were asked !

As far as appearances go, I am an idler, a mere time-killer, without friends, station, or money."

"If you are idle, you will have to amend your ways—"

"And work for you, as well as myself," he interrupted with a laugh.

"As to friends, I would say you could share mine, but then I have so few!—Still—"

"Still, for better or worse you will be Mrs. Gilbert Lisle?"

"Yes,—some day," faltered the young lady.

"I know I am not half as fascinating, nor a quarter as good-looking as Quentin; honestly, what do you see in me, Helen?"

"Do you expect me to pander to your conceit, and to make you pretty speeches?" she asked with a rather saucy smile.

"Indeed, I do not; all the pretty speeches, of course, should come from *me*. I only want to hear the truth," he returned, looking at her with his steady dark eyes.

"Well, then, since you must know, and you seem generally to have your own way, I will try and tell you.—Somehow, from the first—yes, the very *first*—I was sure that you were a person that I could trust; and ever since that time on the wreck—" she paused.

"Yes," he repeated, "ever since that time on the wreck?—go on, Helen."

"I have felt that—that—I would not be afraid to go through anything with you, to —to spend my life with you. *There!*" becoming crimson, she added, "I know I have said too much, *far* too much," clasping her hands together nervously.

A look more eloquent than words illumined Lisle's face.

"And you would give yourself to me in this blind confidence? Helen, I little dreamt when I came down here rather aimlessly, that in these unknown islands, I should find such a pearl beyond price. You cannot understand what it is to me, to feel that I am valued for myself, simply as Gilbert Lisle, poor, obscure, and—" he paused, his voice sounded rather husky, and then he went on, "I must see your father to-night. But how? I left him at billiards. I wonder what he will say to me?"

"Perhaps, perhaps," began Helen rather nervously, "*I* had better speak to him first. I know he likes you, but—"

"Yes, there would seem to be a very considerable *but*," smiling significantly. "Nevertheless, I hope he will listen to me. No, Helen, I would rather talk to him myself."

"At any rate, you will not ask me to leave him for ages,—not for a long time?"

"What do you call a long time?"

"Two or three years; he will be so lonely."

"Two or three years!—and pray what is to become of me?"

"Have you no relations?"

"Yes, some. Chiefly a father, who is pining for the day when I shall introduce him to a daughter-in-law."

"Now you are joking, surely," looking at him with a bewildered face. "I have heard of mothers being anxious to get their daughters married—but a father his sons, never!"

"Ah," repressing a smile, "well, you see, you live and learn."

"And what is your father like?"

" He is old, of course ; he has white hair and a red face, and is short in stature and in temper."

" You do not speak of him very respectfully."

" You are always hauling me up, Helen. First I am lazy, now I am unfilial."

" I beg your pardon. I forget, I am too ready to say the first thing that comes into my head."

" Never mind begging my pardon. I like to be lectured by *you*," taking her hand in his.

" Do not—supposing Mrs. Creery were to see you ? " trying to withdraw hers,—and vainly.

" What if she did ? " he returned boldly ; " it is my own property."

Thus silenced, Helen submitted to have her arm drawn within her lover's, and her hand clasped tightly in his.

" Where does your father live, and what does he do, and like ? " she asked presently.

" He lives in London. What does he do ? Nothing particular. What does he like ?

He likes a rubber of whist, he likes politics, he likes his own way. He is certain to like *you*."

" Oh, I always get on well with old gentlemen," she rejoined with some complacency.

Her companion looked at her with an odd twinkle in his eye, and said,—

" As, for instance ? "

" As, for instance, the General, Colonel Home, Dr. Parkes."

" And you call *them* old gentlemen ! Why, they are men in the prime of life ! Perhaps you consider me an old gentleman also ! "

" Nonsense," she returned with a smile. " Now tell me something about your mother."

" Ah ! my mother," he answered with a sudden change in his expression. " My mother died five years ago."

" I am sorry," began Helen.

" And *I* am sorry, that she did not live to know you. She was the most beautiful woman I ever saw,—and the best."

" You were better off than I was. I do not remember my mother; she was lovely, too," returned Helen, jealous for a certain painted miniature that was the most precious of her treasures.

Mr. Lisle looked at Helen thoughtfully. His mind suddenly travelled back to the night that she had landed on Ross,—and a certain scathing sketch of the late Mrs. Denis. Of course this child beside him, was totally ignorant of her mother's foibles. " The prettiest woman in India " had, at any rate, bequeathed her face to her daughter. Yes, he noted the low brow, straight nose, short upper lip, and rounded chin. But what if Helen had also inherited the disposition of the false, fair, unscrupulous Greek ?

That was impossible ; he was bitterly ashamed of the thought, and mentally hurled it from him with scorn. His lady-love was rather surprised at his long silence ! Of what was he thinking ?

" It is a well-known fact," he said at length, " that the value people place upon themselves is largely discounted by the world ; but when I came down here, merely to see what the place was like, and to shoot and fish, I never guessed, that I should be taken for counterfeit coin by the head of the society for the propagation of scandal."

"Meaning Mrs. Creery," said Helen with a smile.

"Yes. Because I declined to unbosom myself to her, and tell her where I came from, where I was going, what was my age, my religion, &c., &c., she made up her mind that I was a kind of social outcast, and was not to be tolerated in decent company. This, as you may have remarked, sat very lightly on my mind ; I did not come here for society, but it amused me, to see how Mrs. Creery set me down as a loafer and a pauper. It does not always follow, that because a fellow wears a shabby coat, his pockets must be empty.— I am not a poor man ; far from it. Do you think, if I were, I would have the effrontery to go to your father, and say, 'Here I am. I have no profession, no prospects, no money. Hand me over your treasure, your only child, and let us see if what is not enough for one to live on, will suffice for two' ? Were a man to come to *me* with such a suggestion, I should hand him over to the police!"

Helen looked at him in awe-struck astonishment.

" Then you are rich,—and no one guesses it here!"

" Oh, the General knows all about me; so does Quentin; so shall *you!* How I wish," he exclaimed with sudden vehemence, " that these miserable Nicobars had never been discovered! Six weeks will seem a century; especially in the company of Quentin. I shall be obliged to have it out with Master James," he added, with a rather stern curve of his lips. " I had though that lying was an obsolete vice! Only that Hall is going, and is entirely depending on me as a kind of buffer between him and Quentin,—whom he detests,—I would not consider my promise binding. I never knowingly associate with—" he stopped short, and apparently finished the sentence to himself. " Anyway, it will seem years till I come back!"

" And you *will* come back?" she said, looking at him with a strangely wistful face.

For a moment he returned her gaze in reproachful amazement. Then, stretching his hand out towards the east, replied,—

"As sure as the sun will rise there to-morrow, so surely will I return. What have I said or done that you should doubt me now? —you who have trusted me so generously?"

"I cannot tell. I have a strange feeling that I cannot get out of my head; and yet I'm sure you would laugh, were you to hear it, Mr. Lisle."

"Gilbert," he corrected.

"Yes, Gilbert," she repeated softly.

"I must tell you, Helen, what I have more than once been tempted to confide to you. I am not what I seem. I—"

"It was *not* Captain Durand, after all," interrupted a harsh female voice close by, and at this critical moment, Mrs. Creery and Dr. Parkes came swooping down from the hill-top.

"Helen and Mr. Lisle! Well, I declare! Pray do you know that every one is going home? What can you have been thinking of? The band played 'God save the Queen' half an hour ago."

Mr. Lisle drew himself up to his full height (which was five feet ten), and looked as if he wished the good lady—say, at Jericho; and Helen fumbled with her fan, and murmured some incoherent excuse. They both hung back, evidently expecting, and hoping that the elder couple would lead the way down the hill; but, alas! for their expectations, Mrs. Creery suddenly put out a plump hand and drew Helen's reluctant one under her own arm, saying, as she shouldered herself between her and her cavalier,—

" Come along with me ; it's high time little girls like you were at home," and without another word Helen was, as it were, marched off under a strong escort in the direction of the ball-room.

Good-bye to those few transcendental moments, good-bye to the moonlight on the water, the scent of orange-flowers, and all the appropriate surroundings to a love-tale! Say good-bye to Gilbert Lisle and love's young dream, Helen Denis, and go quietly down the hill with Mrs. Creery's heavy arm firmly locked in yours.

The two gentlemen followed in dead silence. Dr. Parkes was infinitely diverted with this little scene; he had been young himself, and it did not need the light of his own past experience to tell him, that this good-looking, impecunious fellow beside him had been trying his hand at making love to the island belle; but Mrs. Creery was a deal too sharp for him, and on the whole, "though he was evidently a gentleman," casting a glance at his companion's aristocratic profile and erect, rather soldierly figure, he considered that it was a deuced piece of cheek for *him* to think of making up to Helen Denis! Alas! little did Dr. Parkes and the careful matron in his van, guess that they were merely carrying away the key of the stable, the steed (meaning the young lady's heart) had been stolen long ago.

As to Mr. Lisle's thoughts, the reader can easily imagine them—disgust, impatience, rage were the least of them. How was he to get another word with Helen? How was he to have a chance of seeing Colonel Denis? Oh! rash and fatal promise that he had made that afternoon. When the ladies all emerged, shawled and cloaked, from the mess-room verandah, he made one bold effort to walk home with his *fiancée;* but every one was leaving simultaneously, and they all descended in one compact body, Dr. Malone escorting Miss Denis on one side, and Captain Rodney on the other; while her accepted lover walked alone behind, and angrily gnawed his moustache. However, he was the last to bid her good-bye, he even went a few paces down the little walk; meanwhile from the high road a crowd looked on—and waited! This was a trying ordeal, and Dr. Parkes' voice was heard shouting impatiently,—

"Now then, Lisle! if you are coming in my boat, look sharp, will you, there's a good fellow?"

He felt a fierce desire to throttle the little doctor! Moments to *him* were more precious than diamonds, and what was half an hour more or less to a dried-up old fogey like that?

He stopped for a second under the palm-trees, and whispered,—

" I'll come over to-morrow early; I mean, this morning, if I may, and if I can possibly manage it; if not, good-bye, darling—our first and last good-bye. I shall be back in six weeks," and then he wrung her hand and went. (A more tender leave-taking was out of the question, in the searching glare of the moonlight, and under the batteries of forty pairs of eyes.)

Poor, ignorant Colonel Denis! who was standing within three yards, little guessed what Gilbert Lisle was whispering to his daughter; indeed, he was not aware that he had been whispering at *all!* nor that here was a robber who wished to carry off his treasure—his all—his one ewe lamb.

No, this guileless, unsuspicious gentleman, nodded a friendly " good-night " to the thief, and went slowly yawning up the steps, then, turning round, said sleepily,—

" Well, and how did my little girl enjoy herself ? "

His little girl looked very lovely in his fond eyes, as she stood below him in her simple white gown, with her face still turned towards the roadway.

"Oh! very, very much, papa!" she replied most truthfully, now entering the dim verandah, and thereby hiding the treacherous blushes that mounted to her very temples.

"That's right!" kissing her as he spoke. "There, be off to bed; it's nearly two o'clock! dreadful hours for an old gentleman like me!"

But Miss Denis did not obey her parent's injunction; on the contrary, she went into the drawing-room, laid down her candle, removed her gloves, and rested her hot face in her hands, and tried to collect her thoughts, and realize her bliss. She was so happy, she could not bear to go to bed, for fear she might go to sleep. She wanted to make the most of the delicious present, to think over every moment, every word, every look, that she had exchanged with Mr. Lisle this most wonderful evening. And to think that all along he had stayed away because he

had thought that she was engaged to Jim
Quentin—he had said so. Jim Quentin!
And she curled her lip scornfully, as she
recollected a recent little scene between that
gentleman and herself.

For a whole hour she sat in the dimly-
lighted drawing-room, looking out on the
stars, listening to the sea, and tasting a
happiness that comes but once in most
people's lifetime. She was rudely aroused
from her mental ecstasy, by a tall figure
appearing in the doorway, clothed in white;
no ghost this—merely her ayah, with her
cloth wrapped round her, saying in a drowsy
voice,—

" Missy never coming to bed to-night ? "

CHAPTER VI.

PROOF POSITIVE.

"About a hoop of gold—a paltry ring that she did
give me." *Merchant of Venice.*

" Is this a prologue—or the poesy of a ring?
'Tis brief, my lord—as woman's love."
Hamlet.

It will not surprise any one to hear, that
there was rather a stormy meeting between
Mr. Lisle and his fellow-inmate. Mr. Quentin did not return home till nearly four
o'clock, and when he did, he found his friend
sitting up for him, and this of itself consti-
tutes an injury, especially when the last-
comer has had rather too much champagne!
Apollo arrived tired and sleepy, with tumbled
locks, and tie, and in a quarrelsome, captious
mood, swearing roundly as he came up the
steps, at his unhappy servants—who had
spent the night in packing.

" Hullo ! " he cried, seeing the other writing at the table, " not gone to roost yet, my early bird ? "

" No," looking at him gravely, " I wanted to speak to you first," rising as he spoke and shutting the door.

" I say ! " with a forced laugh, " you are not going to shoot me, eh ? "

" No, I merely want to ask you why you told me that you were engaged to Miss Denis ? "

" Who says I'm not ? " throwing himself into a chair, and extending his long legs.

" She does," replied his companion laconically.

" And how dare *you* ask her, or meddle in my affairs ? " blustered Mr. Quentin in a loud voice.

" ' Dare ' is a foolish word to use to me, Quentin. I do not want to quarrel with you," feeling that his adversary was not quite himself. " But I wish to know why you deceived me in this way. What was your motive ? "

Mr. Quentin was as much sobered by the

stern eyes of his *vis-à-vis*, as if he had had his head immersed in a bucket of iced water. He reviewed the circumstances with lightning speed; to tide over to-morrow, nay, this very day, was all he wanted. In a few hours they would be off; the *Scotia* sailed at nine, and the chances were ten to one that Lisle and Helen Denis would never meet in this world again. Lisle would probably go home from the Nicobars. He could not afford to get into his black books (for various reasons, chiefly connected with cheque books), and he would brazen it out now. As well be hanged for a sheep as a lamb!

" I *am* engaged to her," he said at last.

" She says you are not; its merely your word against hers."

" And which do you believe? "

" Well, this is no time for mincing matters. I believe Miss Denis," said the other bluntly.

" Believe her against me? A girl you have not spoken to ten times in your life; and you and I have lived here under the same roof like *Brothers* for months. Oh, Gilbert Lisle! " and his beautiful blue eyes looked quite misty, as he apostrophized his companion in a tone as mournful as the renowned " *Et tu, Brute*."—But, as I have already stated, Jim Quentin was a consummate actor.

Mr. Lisle was rather staggered for a moment, and the other went on,—

" Don't you know—but how should you ? for you don't know woman's ways," with a melancholy shake of the head, " that they *all*, even the youngest and simplest of them, think it no harm to tell fibs about their sweethearts ? I give you my solemn word of honour that I've heard an engaged girl swear she was not going to be married to a fellow up to a week before the wedding-day. They think that being known to be engaged, spoils their fun with other men; the more proposals they can boast of the better. If you have been such a fool, as to believe Helen Denis's little joke,—all I can say is, that I am sorry for you ! "

This was hard swearing, certainly, but it was in for a penny, in for a pound, and the *Scotia* sailed at nine o'clock.

Still Mr. Lisle was not convinced, and he saw it and added,—

"You think very little of my bare word, I see. No doubt you would like to see some tangible proof of what I say. There is no time now ('thank goodness,' to himself) to bring us face to face, but if I promise to show you some token before we sail, will that content you?"

Mr. Lisle made no reply.

"And," he continued, "I'm going to turn in now, for its four o'clock, and I'm dead beat. Don't let us fall out, old fellow—no woman is worth it. They are all the same, they can't help their nature," and with this parting declaration Mr. Quentin, finished actor and finished flirt, sorrowfully nodded his head and took his departure.

Once in his own apartment, he tore off his coat, called his body servant to pull off his boots, threw himself into an arm-chair, and composed himself with a cheroot, yea, at four o'clock in the morning! He had shown a bold front, and had impressed Lisle—that he could see plainly. But how about this little token? He did not possess a glove, a ribbon, a flower, much less a photograph or a lock of hair. What was he to do? For fully a quarter of an hour the query found no answer in his brain, till his sleepy servant, asking some trivial question, gave him a clue; he saw it all, as it were, in a lightning flash.

Abdul was married to Miss Denis's ayah (a handsome, good-for-nothing virago, who, it was rumoured, occasionally inflicted corporal punishment upon her lord and master, and was avaricious to the last degree).

Abdul was a dark, oily-looking, sly person, who was generally to be trusted—when his own interests did not clash with his employer's.

"Abdul, look here," said Mr. Quentin suddenly, "I want you to do something for me at once."

"Yes, saar," said Abdul in a drowsy voice.

"Go off, now, this moment, and get the boat, go across to Ross"—here Abdul's face became very blank indeed,—"go to

Colonel Denis's bungalow, and speak to Fatima, and tell her." Mr. Quentin was, for once in his life, a little ashamed of what he was about to do; but do it he would, all the same—he *must*—he had burnt his boats. "Tell her to give you that queer gold ring Missy wears—no stones, a pattern like this," talking the jargon of the East, and showing an ancient seal. "I want it as "muster" for another, just to look at; for a present for Missy, and will give it back to-day. Mind you, Abdul, never letting Missy know : if you do, or if Fatima says one word, you get nothing; if you and she manage the job well, you shall have twenty rupees ! "

Abdul stared, and then salaamed and stolidly replied,—

" I never telling master's business, master knows."

" Then be off at once, and let me see you back by seven o'clock ; and don't attempt to show your face without *that*, or no rupees—you understand ? "

" Master pleases," ejaculated Abdul, and

vanished on his errand, an errand that was much to his taste. A little mystery or intrigue, and the prospects of a good many rupees, appeals to the native mind in a very direct fashion.

At seven o'clock he had returned, having accomplished his mission. Breathless and radiant he appeared, and roused his sleeping master, saying,—

"I've come back, saar, and here"—unfolding a bit of his turban, and holding out his hand—"I've brought the pattern master wanted."

"By Jove!" leaning up on his elbow, and now wide awake, "so you have," taking Helen's ring, and surveying it critically. Yes! nothing could be better; she always wore it on the third finger of her right hand, and there was surely some history about it, or he was much mistaken. "We will see what Lisle will say to *this*," he muttered to himself as he squeezed it on his own somewhat plump little finger. Then to Abdul,—

"Very well. All right; I'll give it back, you know. Meanwhile go to my box over

there, and bring the money-bag, and
count yourself out the dibs I promised
you."

Abdul obeyed this order with great alacrity,
salaamed, and then waited for his next
instructions.

"You can go now; call me in half an
hour," said his master, dismissing him with
a wave of his newly-decorated hand.

"A first-class idea! and, by Jove, Miss
Helen, I owed you this. The idea of a little
chit like you, the penniless daughter of an
old Hindoo colonel, giving yourself such airs
as you did last night," alluding to a scene
when Helen, wearied by his compliment, and
indignant at his presumption, had plucked up
courage to rebuke him in a manner that
penetrated even the triple armour of his self-
conceit. Such a thing was a novel experience,
the recollection of it stung him still, and
to such a man as Jim Quentin, the affront
was unpardonable. It awoke a slumbering
flame of resentment in his rather stolid
breast, and a burning desire to pay her out!
And he would take right good care that she

did not catch Lisle—Lisle who was certainly inclined to make an ass of himself
about her. With this determination in his
mind, he rose, dressed, and languidly lounged
into their mutual sitting-room, where his
companion had been impatiently awaiting
him for an hour, intending subsequently to
sail across to Ross, and take one more
parting with his fair lady-love, and, if possible,
obtain a word with her father.

"So you have appeared at last?" he
exclaimed; "I've been expecting you for
ages."

"Have you? but we need not leave this
till half-past eight," looking at his watch.
"They know we are going,—and Hall is
never in time."

"I'm not thinking of the *Scotia*," returned the other, scarcely able to restrain
his impatience; "but of what you promised
to show me last night—that proof you spoke
of, you know."

"Oh! yes; by-the-bye, so I did," as if
it were a matter of the most complete indifference. "I daresay I have something that

will convince you. Will this do?" tendering his hand as he spoke, in quite an airy,
nonchalant fashion.

Mr. Lisle glanced at it, and beheld his ring,
the wreck ring, adorning Jim Quentin's
little finger! He started as if he had been
struck—his own gift, that she declared
she would never part with! And she had
bestowed it already,—given it to Quentin:
this was enough, was too much—he asked
no more.

"Well, will that do?" demanded Apollo,
removing and tendering the token. "Are
you satisfied *now?*"

"Yes," replied Mr. Lisle, who had re-
gained his self-command. But the other had
noted the sudden pallor of his face, the
almost incredulous expression of his eyes,
and felt that this borrowed bit of jewellery
was indeed a trump card, boldly played.

Jim was immensely relieved as this one
syllable fell from his companion's lips. The
whole matter was now settled. Lisle was
choked off: his own credit was unimpeached,
but it had had a narrow squeak, and last

night he had undoubtedly spent a very unpleasant quarter of an hour.

Of course Mr. Lisle did not return to
Ross, although the white boat lay waiting
for him for an hour, by the landing steps.
Helen had more than half expected him,
with trembling, delightful anticipations; how
many times did she run to look in the glass?
how many times re-arrange the flowers in
her dress? how many times did she dart to
the verandah as a manly step came up the
road? But, alas! after an hour's expectation,
her hopes were dashed to the ground by
Miss Lizzie Caggett.

" The *Scotia* has sailed!" she screamed out
from the pathway. " Come up to the flagstaff, and see the last of her."

It was the custom for the ladies on Ross
to take constitutionals before breakfast, and
Helen, on her way to the top of the hill with
Miss Lizzie, was joined by Mrs. Creery, Mrs.
Home, and Mrs. Durand, all discussing the
previous evening's dissipation. Helen was
(they all remarked) unusually silent: generally she was full of fun and spirits. She

stood aloof, looking after the receding steamer, and said to herself, "What if he should never come back!"

But this was a merely passing thought that she silenced immediately. Mr. Lisle was, as every one knew, a man of his word, and never broke a promise.

The little group of ladies stood watching the smoke of the steamer become smaller and smaller till it vanished altogether, and Helen, as she turned her face away from the sea at last, had a suspicion of tears in her eyes,— tears which her companions attributed to Mr. Quentin. As she walked down the hill with Mrs. Home, that warm-hearted little lady, who was leaning on her, pressed her arm in token of sympathy, and whispered in a significant tone,—

"He will come back, dear."

"So he will," agreed Helen, also in a whisper, blushing scarlet as she spoke. But she and Mrs. Home were not thinking of the same person!

CHAPTER VII.

"A GREAT BATTLE."

"But 'twas a famous victory."

Southey.

IT is perhaps needless to mention that Mrs. Creery made it her business, and considered it her duty, to circulate the intelligence that she had received about Mr. Lisle without unnecessary delay. She read portions of the letter referring to him, in "strict confidence," to every one she could get hold of, and the missive was nearly worn out from constant folding and unfolding. If any one ventured to impugn her testimony, she would lay her hand upon her pocket with a dramatic gesture, and say,—

"That's nonsense! I've got it all here in black and white. I always knew that there was a screw loose about that man. Perhaps you will all be guided by *me* another time!

I'm an excellent judge of character, as my
sister, Lady Grubb, declares. She always
says, 'You cannot go far wrong if you listen
to Eliza'—that's me," pointing to her breast
bone with a plump forefinger. Then she
would produce the billet and, after much
clearing of throat, commence to read what
she already knew by heart.

" 'You ask me if I can tell you anything
about a Mr. Lisle, a mysterious person who
has lately come to the Andamans; very
dark, age over thirty, slight in figure, shabby
and idle, close about himself, and with a
curious, deliberate way of speaking; supposed to have been in the army, and to have
come from Bengal. Christian name unknown,
initial letter G.' "

(It sounded exactly like a description in
a police notice.)

" 'My dear Mrs. Creery, I know him well,
and he may well be close about himself and
his affairs' "—here it was Mrs. Creery's cue
to pause and smack her lips with unction.
" 'If he is the person you so accurately de-
scribe, he is a Captain Lisle, a black sheep

who was turned out of a regiment in Bengal
on account of some very shady transactions on
the turf.' He told me himself he was fond
of riding," Mrs. Creery would supplement, as
if this fact clinched the business. " ' He was
bankrupt, and had a fearful notoriety in
every way. No woman who respected herself would be seen speaking to him! The
Andamans, no doubt, suit him very well at
present, and offer him a new field for his
energies, and a harbour of refuge at the same
time. Do not let any one cash a cheque for
him, and warn all the young ladies in the
settlement that he is a *married* man ! ' "

"There," Mrs. Creery would conclude,
with a toss of her topee, "what do you
think of that ?"

" Mr. Lisle is not here to speak for himself," ventured Helen on one occasion. " *Les
absens ont toujours tort.*"

It was new to see Helen adopt an insurrec-
tionary attitude. Mrs. Creery stared.

" Nonsense—stuff and nonsense," angrily.
" And let me tell you, Helen Denis, that it is
not at all maidenly or modest for a young

girl like you to be taking up the cudgels for a notorious reprobate like this Lisle."

"I'm sure he is not a reprobate, and I'm certain you are mistaken," rejoined Helen bravely.

Here the elder lady flamed out, and thumped her umbrella violently on the ground, and cried in her highest key,—

" Then why did he go away ? He knew that I had heard about him, for I told him so to his face. I never say behind a person's back what I won't say to their face." (Oh ! Mrs. Creery, Mrs. Creery !) "And it is a very remarkable coincidence, that in less than twelve hours, he was out of the place ! How do you account for that, eh ? "

She paused for breath, and once more proceeded triumphantly,—

" He will never show here again, believe me ; and, after all, I am thankful to say he has done no great harm ! As far as *I* know he ran no bills in the bazaar, and certainly neither you nor Lizzie Caggett lost your hearts to him ! "

Helen became very pale, her lips quivered,

and she was unable to reply for a moment. Then she said,—

"At any rate, I believe in him, Mrs. Creery,—and always will; deeds are better than words. Have you forgotten the wreck?"

"Forgotten it?" she screamed. "Am I ever likely to get it out of my head? Only for my calling myself hoarse, you and Mr. Lisle would both have been murdered in that hole of a cabin! You know I told you not to go down, and you would, and see what you got by it."

There was not the slightest use in arguing with this lady, who not only imposed upon others, but also upon herself; she had a distorted mind, that idealized everything connected with her own actions, and deprecated, and belittled, the deeds of other people! The only persons who had *not* heard the horrible tale about Mr. Lisle were the Durands and the general; the latter was a singularly astute gentleman, and never lost a certain habit of cool military promptitude, even when in retreat. Each time Mrs. Creery

had exhibited symptoms of extracting a letter
from her pocket, he had escaped! The
Durands were Mr. Lisle's friends,—a fact that
lowered them many fathoms in Mrs. Creery's
estimation, and were consequently the very
last to hear of the scandal!

About a fortnight after the departure of
the *Scotia*, the general gave one of his usual
large dinner-parties; every one in Ross was
invited, and about twenty-four sat down to
the table. When the meal was over, and the
ladies had pulled a few crackers, and sipped
their glass of claret, they all filed off into
the drawing-room in answer to Mrs. Creery's
rather dramatic signal, and there they looked
over photographs, noted the alterations in
each other's dresses, drank coffee, and conversed in groups. In due time the conversation turned upon that ever fertile topic, "Mr.
Lisle," and Mrs. Graham, who was seated
beside Mrs. Durand, little knowing what she
was doing, fired the first shot, by regretting
very much "that Mr. Lisle had turned out
to be such a dreadful character, so utterly
different from what he seemed." Encouraged

by one or two cleverly-put questions from her neighbour, she unfolded the whole story. Meantime, Mrs. Durand sat and listened, in rigid silence, her lips pressed firmly together, her hands lightly locked in her pale-blue satin lap. When the recital had come to an end, she turned her grave eyes on her companion, and said in her most impressive manner,—

" *How* do you know this ? "

" Oh, it's well known, it's all over the place. Mrs. Creery had a letter," glancing over to where that lady reclined in a comfortable chair, with a serene expression on her face, and a gently-nodding diadem.

"Mrs. Creery," said Mrs. Durand, raising her voice, which was singularly clear and penetrating, " pray what is this story that you have been telling every one about Mr. Lisle ? "

This warlike invocation awoke the good lady from her doze, and, like a battle-steed, she lifted her head, and, as it were, sniffed the conflict from afar !

" I've been telling nothing but the truth,

Mrs. Durand "—rousing herself at once to an
upright position—"and you are most welcome to *hear* it, though he *is* a friend of
yours," and she tossed her diadem as much
as to say " Come on ! "

" Thank you!   Then will you be so very
kind as to repeat what you have heard,"
returned Mrs. Durand with a freezing polite-
ness that made the other ladies look at each
other significantly.   There was going to be a
fight, and they felt a thrill of mingled delight
and apprehension at the prospect.

Bold Mrs. Durand was the only woman in
the island who had never veiled her crest to
Mrs. Creery.   She was now about to chal-
lenge her to single combat—yes, they all saw
it in her face !

" I always knew that there was something
very wrong about that man," began the elder
lady in her usual formula, and figuratively
placing her lance in rest.   " People who
have nothing to hide, are never ashamed to
speak of their concerns, but no one ever got
a word out of Mr. Lisle, and I am sure he
received every encouragement to be open ! He

was in the army, he admitted *that* against his will, and that was all. He never deceived *me ;*—I knew he was without any resources, I—knew he was out at elbows, I knew—"

"Pray spare us your opinion, and tell us what *facts* you have to go upon," interrupted Mrs. Durand, calmly cutting short this flow of denunciation.

"I have a letter from a friend at Simla," unconsciously seeking her pocket, " a letter," she retorted proudly, " which you can *read,* saying that he was cashiered for conduct unbecoming an officer and a gentleman, that he is a bankrupt, and a swindler, and a married man," as if this last enormity crowned all.

"It is not true—not a word of it ! " replied Mrs. Durand, as coolly as if she were merely saying, "How do you do."

" Not true ! nonsense ; is he not dark, aged over thirty, name Lisle ? did he not hang about the settlement for six months living on his wits ? of course it is true," rejoined the elder lady, with an air that proclaimed that she had not merely crushed, but pulverized, her foe!

"Lisle is not an uncommon name, and I know that my friend is not the original of your flattering little sketch."

"But I tell you that he *is!* I can prove it; I have it all in black and white!" cried Mrs. Creery furiously—her temper had now gone by the board. Who was this Mrs. Durand that she should dare to contradict her? She saw that they were face to face in the lists, and that the other ladies were eager spectators of the tourney; it was not merely a dispute over Mr. Lisle, it was a struggle for the social throne, whoever conquered now, would be mistress of the realm. This woman must be brow-beaten, silenced, and figuratively slain!

"I have it all in writing, and pray what can *you* bring against that?" she demanded imperiously.

"Simply my word, which I hope will stand good," returned the other firmly.

Mrs. Creery laughed derisively, and tossed her head, and then replied,—

" Words go for nothing ! "

This was rude—it was more than rude, it was insulting !

" Am I to understand that you do not believe mine ? " said Mrs. Durand, making a noble effort to keep her temper.

" Oh," ignoring the question, " I have never doubted that *you* could tell us more about Mr. Lisle than most people ! and a woman will say anything for a man—a man who is a friend," returned the other lady with terrible significance.

This was hard-hitting with a vengeance, still Mrs. Durand never quailed.

" Shall I tell you who Mr. Lisle really is ? I did not intend to mention it, as he begged me to be silent."

(Here Mrs. Creery's smile was really worth going a quarter of a mile to see.)

" I have known him for many years ; he is an old friend of mine, and of my brothers."

" Oh, of your brothers ! " interrupted her antagonist, looking up at the ceiling with a derisive laugh, and an adequate expression of incredulity.

" I am not specially addressing myself to *you*, Mrs. Creery," exclaimed Mrs. Durand at white heat, but still retaining wonderful command of her temper. " My brothers were at Eton with him," she continued, looking towards her other listeners. " He is the second son of Lord Lingard, and the Honourable Gilbert Lisle."

A silence ensued, during which you might have heard a pin drop; Mrs. Creery's face became of a dull beetroot colour, and her eyes looked as if they were about to take leave of their sockets.

" And what brought him masquerading here ? " she panted forth at last.

" He was not masquerading, he came in his own name," returned Mrs. Durand with calm decision. " He left the service on coming in for a large property, and spends most of his time travelling about; he is fond "—addressing herself specially to the other ladies, and rather wondering at Helen Denis's scarlet cheeks—" of exploring out-of-the-way places. I believe he has been to Siberia, and Central America. The Andamans were a novelty; he came for a few weeks, and stayed for a few months, because he liked the fishing and boating and the unconventional life."

" And who is the other Lisle ? "

" Some distant connection, I believe ; every family has its black sheep."

" Why did he not let us know his position ? " gasped Mrs. Creery.

" Because he thinks it of so little importance ; he wished, I conclude, to stand on his own merits, and to be valued for himself alone. He found his proper level here, did he not, Mrs. Creery ? he lived in the palace of truth for once ! " and she laughed significantly—undoubtedly turn-about is fair play, —it was her turn now.

" I must say that I wonder what he saw in the Andamans," exclaimed Mrs. Grahame at last.

" One attraction, no doubt, was, because he could go away whenever he liked; another, that he was left to himself—no one ran after him ! " and Mrs. Durand laughed again. " In London he is made so much of ; as every one knows he is wealthy, and a bachelor, and that his eldest brother has only one lung ! Besides all these advantages, he is extremely popular, and is beset by invitations to shoot, to dance, to dine, to yacht, from year's end to year's end. Well, he got a complete holiday from all that kind of thing *here !* "

Then she recollected that in castigating Mrs. Creery and Miss Caggett she was including totally innocent people—people who had always been civil to the Honourable Gilbert Lisle, such as Mrs. Grahame, Mrs. Home, Miss Denis, and others, and she added,—

" All the same, I should tell you that he enjoyed his stay here immensely, he told me so, and that he would always have a kindly recollection of Port Blair, and of the friends he had made in the settlement."

(Mrs. Durand, thought Helen, does not know everything ; she evidently is not aware that he is coming back !) The speaker paused at the word settlement, for she had made the discovery, that most of the gentlemen had entered, and were standing in the background, while she had been, as it were, addressing the house! A general impression had been gathered about Mr. Lisle also, as Captain Rodney whispered to Dr. Malone, that "Mrs. Creery had evidently had what she would be all the better for, viz. a rare good setting down."

Infatuated Mrs. Creery! deposed, and humbled potentate! if there was one thing that was even nearer to her heart than Nip, it was the owner of a *title*.

She could hardly grasp any tangible idea just at present, she was so completely dazed. It was as if Mrs. Durand had let off a catherine-wheel in her face.

Mr. Lisle an Honourable! Mr. Lisle immensely rich! Mr. Lisle, whom she had offered to pay for his photographs, whom she had never met without severely snubbing! And all the time, he was the son of a lord, and she had unconsciously lost a matchless opportunity, of cementing a life-long friendship with one of the aristocracy! Alas, for poor Mrs. Creery! her mind was chaos!

After the storm, there ensued the proverbial calm; the piano was opened, and
people tried to look at ease, and to pretend,
forsooth! that they were not thinking of the
recent grand engagement, but it was all a
hollow sham.

Helen, if it had been in her power, would
have endowed that brave woman, Mrs. Durand,
with a Victoria Cross for valour, and, indeed,
every lady present, secretly offered her a
personal meed of admiration and gratitude.
She had slain their dragon, who would never
more dare to rear her head, and tyrannize
over the present, or vilify the absent. Surely
there should be some kind of domestic decoration, accorded to those, who arm themselves
with moral courage, and go forth and rescue
the reputation of their friends!

Miss Caggett sat in the background, looking unusually grave and gloomy—no doubt
thinking with remorseful stings of *her* lost
opportunities. Dr. Malone grinned and
nodded, and rubbed his rather large bony
hands ecstatically, and whispered to Captain
Rodney that "*he* had always had a notion,

that Lisle the photographer, was a prince in disguise!"

As for Mrs. Creery, as before mentioned, that truculent lady was absolutely shattered; she resembled an ill-constructed automaton who had been knocked down, and then set up limply in a chair, or a woman in a dream,—and that a bad one. After a while she spoke in a strangely subdued voice, and said,—

"General, I don't feel very well; that coffee or yours has given me a terrible headache! If you will send for my jampan, I'll just go quietly home."

Thus she withdrew, with a pitiable remnant of her former dignity, her host escorting her politely to the entrance, and placing her in her chair with faint regrets. Every one knew perfectly well, that it was *not* the general's coffee that had routed Mrs. Creery, it was she whose beautiful contralto was now filling the drawing-room, as her late antagonist tottered down the steps,—it was that valiant lady, Mrs. Durand!

CHAPTER VIII.

THE NICOBARS.

"Once I loved a maiden fair,
    But she did deceive me."

WHEN last we saw Mr. Quentin, he had just
succeeded in convincing his companion that
he was Miss Denis's favoured suitor. This
was well,—this was satisfactory. But it was
neither well, nor yet satisfactory, to behold
Lisle calmly appropriate the posy ring, and
put it in his waistcoat pocket.

"Hullo! I say, you know," expostulated
Apollo, "give me back my property!"

"No," returned the other very coolly; "it
was originally mine, and as it has once more
come into my hands, I will keep it."

Mr. Quentin became crimson with anger
and dismay.

"I found it on the wreck, and gave it to
Miss Denis, who said she valued it greatly,

but as she has passed it on to you, I see
that her words were a mere *façon de parler*,
and if she asks you what you have done with
it? you can tell her that you showed it to
me,—and that *I* retained it."

There was a high-handed air about this
bare-faced robbery that simply took Mr.
Quentin's breath away, and the whole pro-
ceeding put him in, as he expressed it himself, "such an awful hat," for he had never
meant to steal the ring, he only wanted the
loan of it for half an hour, and now that it
had served his purpose, it was to be restored
to its mistress; but here was Lisle actually compelling him to be a *thief!* Vainly
he stammered, blustered, and figuratively
flapped his wings! he might as well have
stammered, and blustered, to the wall.
Lisle was impassive,—moreover, the boat was
waiting; and Abdul returned to Ross and
Fatima, plus twenty rupees, but minus the
ring.   And what a search there was for that
article when Helen Denis missed it; rooms
were turned out, matting was taken up,
every hole and corner was searched, but

all to no purpose,—considering that the
ring was, as we know, on its way to the
Nicobars.

Fatima, the Cleopatra-like, was touched,
when she saw her missy actually weeping
for her lost property; but all the same, she
positively assured her, that she had never
seen it since she had had it on her finger
last—indeed, if it had been in her power to
return it, she would have done so, for Helen
offered a considerable reward to whoever
would restore her, the most precious of her
possessions. Days and weeks went by, but
no ring was found.

The *Scotia* left Calcutta once every six
weeks, calling firstly at Port Blair, then at
the Nicobars, then Rangoon, and so back to
Calcutta; and the reason of Mr. Quentin's
hurried departure was, that the order to
start for the Nicobars, came in the steamer
that was to take him there, otherwise there
would have been the usual delay of six weeks.
Once on board, he went straight below to his
cabin, turned in, and recouped himself for his
sleepless night. He slept soundly all day long,

having immense capacities in that line. Mr.
Hall, the settlement officer, walked the
deck with Mr. Lisle, and subsequently they
descended to the saloon, and played chess.
The group near the flagstaff had not been
unnoticed by the passengers of the *Scotia*
as she steamed by under the hill; there had
been some waving of handkerchiefs, but Mr.
Lisle's had never left his pocket—he had something else in that self-same pocket, that for-
bade such demonstration—the fatal ring, and
a ring, that bore for motto, as he had now
discovered, " Love me and leave me not "
(a motto that implied a bitter mockery of
the present occasion). This wreck ring was
assuredly an unlucky token ! Only last night,
and Helen had seemed to him, the very incar-
nation of simplicity, truth, and faith—what a
contrast to those many lovely London Sirens
who smiled on him—and his *rent roll !*
Never again would he be deceived by nine-
teen summers, and sweet grey eyes—no,
never again—this was the determination he
came to, as he paced the deck that night
beneath the stars.

The next morning the *Scotia* was off the low, long coast of the Nicobars; so low was it, that it resembled a forest standing in the water. In the midst of this seeming forest, there was a narrow passage, that a casual eye might easily overlook; a passage just barely wide enough to admit the steamer, with a natural arch of rock on one side; the water was clear, emerald green, and very deep, and along the wooded shores of the entrance to Camorta were many white native huts, built on wooden piles, scattered up and down the high banks clothed in jungle. Soon the passage widened into a large inland bay, lined with mangroves, and poison-breathing jungles, save for a clearing on the left hand side, where there was a rude pier, a bazaar of native houses, and some larger wooden buildings on the over-hanging hill—this was Camorta, the capital of the Nicobars, to which Port Blair was as London to some small provincial town.

The natives were totally different to the Andamanese; they were Malays, with brown skins, flat heads, and wide mouths, and came swarming round the three Europeans as they landed, and commenced to climb the hill. One, who was very sprucely dressed in a blue frock-coat, grey trousers, white tie, and tall hat, and flourished a gold watch, was barefooted, and had it made known to Mr. Lisle, before he was five minutes on *terra firma*, that he was prepared to give him one thousand cocoanuts, in exchange for his boots! .

The buildings on the hill, included a big, gaunt-looking bungalow, in which the three new arrivals took up their quarters. It was rather destitute of furniture, but commanded a matchless view of this great inland bay and far-away hills; it also overlooked a rather suggestive object, an old white ship, that lay off Camorta, the crew of which had been killed and eaten, many years previously, by the inhospitable Nicobarese! Gilbert Lisle, had never in all his wanderings, been in any place he detested as cordially as his present residence. Days seemed endless, the nights hot and stifling, the sun scorching, the sport bad. And other things, such

probably as his own frame of mind did not
tend to enhance the charms of Camorta.
Mr. Hall had ample occupation, Jim Quentin
an unlimited capacity for sleep.   He had also
a box full of literature, a good brand of
cigars, and, moreover, was at peace with
himself, and all mankind.   He could do a
number of doubtful actions, and yet he always
managed to retain himself in his own good
graces.   He had squared Lisle, who was
going away direct from the Nicobars to
Rangoon, thence to Singapore and Japan; this
was a most desirable move, and there would
be no more raking up of awkward subjects,
and *he* would never be found out.   His period
of expatriation was nearly at an end, he was
financially the better for his exile at Port
Blair, and then, hurrah for a hill-station,
fresh fields, and pretty faces, or, better still,
Piccadilly and the Park!   Meanwhile, he was
at the Nicobars, and there he had to stay, so
he accepted the present philosophically, and
slept as much as possible, and grumbled
when awake, at the food, the climate, and the
heads of his department, and was not nearly

as much to be pitied as he imagined, not half as much as Lisle, who neither read novels nor slept many hours at a stretch, or had agreeable anticipations of future flirtations in hill-stations. He was remarkably silent, and smoked many of the drowsy hours away. When he *did* join in the conversation, his remarks were so cynical, and his words so sharply edged that Mr. Quentin was positively in awe of him,—and was more than usually wary, in the choice of his topics. Out of doors, he shot the ugly, greedy caymen, caught turtle, and sketched, or explored the country recklessly ; making his way through the rank, dank, jungle, where matted creepers hung from tree to tree, and snakes and spotted vipers, darted up their hideous heads, as he brushed past their moist, dark, hiding-places.

A good deal of Mr. Lisle's time was spent, in absolute idleness, and though the name of Helen Denis never crossed his lips, he had by no means cast her out of his mind.—Hourly he fought with his thoughts, hourly he weighed all the *pros* and *cons*. Her accept-

ance of Quentin's attentions, went to balance
against her coolness to him subsequently;
her blushes, when he appeared, were a set-off
against her solemn denial of any understanding between them ; her evident agitation when he himself had wooed her, was
neutralized by the bestowal of his ring upon
Quentin—the ring kicked the beam, the
ring was the verdict! After all, Quentin was
ten times more likely to engage a girl's fancy
than himself. Apollo was handsome, gay,
and fascinating, (when he chose ;) *he* was sunburnt, shabby, rather morose, and seemingly
a pauper—that part of it was his own fault,
he had no one but himself to blame for that.
Query, would it have been better if he had
permitted the truth to leak out, and allowed
the community to know, that they had the
Honourable Gilbert Lisle, the owner of ten
thousand a year, dwelling among them ? In
some ways, things would have been pleasanter, but he had not come down to the
Andamans for society, but for sea-fishing,
and sailing, and an unfettered, out-door life.
And when he was accidentally thrown into

the company of a pretty girl, who was as pleasant to him as if he were a millionaire, who smiled on him as brightly as on others, in far more flourishing circumstances, who could ask him to resist the temptation that had thrust itself into his way—the triumph of winning her in the guise of a poor, and unpretending suitor?

The temptation led him on, and dazzled him, and for a moment he seemed to have the prize in his hands; and what a prize! especially to him, who was accustomed to being flattered, deferred to, and courted in a manner that accounted for his rather cynical views of society. But, alas! his treasure-trove (his simple-minded island maiden), had been rudely wrested from him ere he had realized its possession; and yet, after all, it was no loss, the apparently priceless jewel was imitation, was paste!

Why had she told him a deliberate lie? He might forgive a little coquetry, (perhaps;) he might forgive the unpleasant fact of her having " made a fool of him," as his friend had so delicately suggested, but a falsehood, uttered without a falter or a blush, *never !*

Week succeeded week, and each day seemed as long as seven—each week a month. Lisle, the ardent admirer of strange scenes, and strange countries, was callous and indifferent to the natural beauties of the place. He had actually come to *hate* the magnificent foliage, golden midday hazes, and the gorgeous, blinding sunsets, of these sleepy southern islands. All he craved for, was to get away from such sights, and never, never, see them more ! Latterly, he found ample occupation in nursing Mr. Hall to the best of his ability. Mr. Hall, who had fallen a victim to the deadly Nicobar fever, and tossed and moaned and raved all through the scorching days and suffocating nights, and was under the delusion that the hand that smoothed his pillow, and held the cup to his parched lips, and bathed his burning temples, was his mother's ! Jim Quentin (the selfish), merely contented himself with languidly inquiring after the patient once a day, and shutting himself up in his own side of the bungalow, as it were in a fastness,

K 2

partaking of his meals alone, totally ignoring his companions, since one of them was sick, and the other was stupid.

The thin veneer of Mr. Jim's charm of manner, could not stand much knocking about; a good deal of it had worn off, and Mr. Lisle beheld him, as he really was ; selfish to the core, vain, and arrogant,—yet not proud, not very sensitive on the subject of borrowing money, and with rather hazy ideas, with regard to the interpretation of the word " honour."

Lisle, in his heart, secretly despised his fascinating inmate ; but, needless to say, he endeavoured to keep this sentiment entirely in the background, though, now and then, a winged word, like a straw, might have shown a looker-on which way the wind blew.

At length the long-desired *Scotia* came steaming up Camorta Bay, like a gaoler to set free her prisoners ; she remained off the pier for a few hours, and Mr. Lisle was unfeignedly delighted to see her once more, for she was to carry him away to Rangoon, to civilization, occupation, and oblivion.   His traps were ready, but ere he took leave of his companions and went on board, he sat for a while reading the newly-arrived letters in the verandah, along with Jim Quentin.

"Hullo!" exclaimed the latter, suddenly looking up. "I say, what do you think! here is a letter from Parkes, and poor old Denis is dead!"

"Dead?" ejaculated his companion.

"Yes, listen to this,"—reading aloud,— "he was on the ranges one morning, and in trying to save a native child who ran across the line of fire, he was shot through the heart. We are all very much cut up, and as to Miss Denis, the poor girl is so utterly broken down, you would scarcely know her."

"It must have been a fearful shock," said Mr. Lisle. "I'm very sorry for Denis, very. Of course you will go back at once—now!"

"How?" thrown completely off his guard, "why?"

"How? by the *Enterprise*, which will be here in three days with stores, and why? really, I scarcely expected you to ask *me* such a question. She—"

"Oh," interrupting quickly, "oh, yes! I quite understand what you mean. Oh, of course, of course!"

After this ensued a rather long silence, and then Mr. Lisle spoke,—

"I now remember rather a strange thing," he said reflectively. "Denis and I were looking over the wall of the new cemetery together one evening, and I recollect his saying, that he wondered how long it would be till the first grave was dug.—Strange that it should be his own!"

"Strange indeed!" acquiesced his companion tranquilly, "but of course, everything must have a beginning. Here's a Lascar coming up from the pier," he added, rising hastily, and collecting his letters as he spoke, "and we had better be making a start."

In another hour Mr. James Quentin was walking back to the Bungalow alone. As he stood on the hill above the pier, and watched the smoke of the departing steamer above the jungle, he felt a curious and unusual sensation, he actually felt,—his almost fossilized conscience told him,—that he had not behaved altogether well to Lisle! Lisle, who had been his friend by deeds, not words; Lisle, who had borne the blow he had dealt him like a man; had never once allowed a word, or allusion that might reflect on Helen, to pass his lips, and had accepted the ring with unquestioning faith. Yes, Lisle, though rather silent and unusually dull, (for generally he was such an amusing fellow,) had taken his disappointment well. Mr. Quentin, however, rated such disappointments very lightly. Judging others by himself, they were mere pin-pricks at the time, and as such consigned to the limbo of complete oblivion within a week.

"After all," he said aloud, as he slowly strolled back with his hands in his pockets, "I am in reality his *best* friend! It would never have done for him, to entangle himself with a girl without connections, a girl without a penny, a girl he picked up at the Andamans! Haw, haw! by Jove! how people would laugh! No, no, Gilbert Lisle, you must do better than that; you will have to look a little higher for the future Lady Lingard. I don't suppose she has a brass farthing, and she certainly would not suit my book at all—"

Needless to add, that this mirror of chivalry did not return to Port Blair an hour sooner than was his original intention.

CHAPTER IX.

THE FIRST GRAVE.

"They laid him by the pleasant shore,
And in the hearing of the wave."

Tennyson.

THE news about Colonel Denis was only too
true! He had started for the ranges on
Aberdeen one morning about nine o'clock,
as his regiment was going through their
annual course of musketry, and as he stood
in a marker's butt, close to the targets, a
native child from the Sepoy lines, suddenly
emerged from some unsuspected hiding-place,
where she had been lying *perdue*, and ran
right into the open, across the line of fire.
Colonel Denis rushed out to drag her into
shelter, but just as he seized her, a bullet
from a Martini-Henry struck him between
the shoulders, and without a groan, he fell

forward on his face, dead.   Yes, he was quite dead when they hurried up to him.

The shock to every one was stupefying; they were speechless with horror; but five minutes previously he had been talking to them so cheerfully, and had to all appearances as good a life as any one present,—and now here he lay motionless on his face in the sand, a dark stain widening on his white coat, and a frightened little native child whimpering beside him.

"Instantaneous," said Dr. Malone, with an unprofessional huskiness in his voice, when they brought him running to the spot. "What an awful thing, and no one to blame, unless that little beggar's mother," glancing at the imp, who stared back at the sahib with all the power of her frightened black eyes. "Poor Denis; but it was just like him,—he never thought of himself." This was his epitaph, the manner in which he met his death, 'was just like him.'"

And who was to break the terrible tidings to his daughter? People asked one another the question with bated breath, and anxious

eyes, as they stood around. Who was to go
and tell her, that her father, to whom she
had bidden a playful good-bye an hour ago,
was dead, that that smiling wave of his hand
had been, Farewell for ever !

. . . .

It was about eleven o'clock, and Helen
was sitting at the piano, playing snatches
of different things, unable to settle down to
any special song or piece. She had felt
curiously restless all the morning, and was
thinking that she would run over, and have
a chat with Mrs. Home,—for she was too idle
to do anything else,—when a sudden loud
sob, made her start up from the music-stool
and turn round somewhat nervously.

There she beheld her ayah, Fatima, staring
at her through the purdah, but the instant
she was discovered, she quickly dropped it,
and vanished. It never occurred to Helen
to connect Fatima's tears with herself, or her
affairs ; it was more than probable that she
had been having a quarrel with her husband,
and that they had been beating one another,
as was their wont,—when words were ex-

hausted !   She was thinking of following her
handmaiden, but she believed it would only be
the old story, "Abdul, plenty bad man, very
wicked rascal," when her ear caught the sound
of footsteps coming up the front pathway.
They halted, then it was *not* Mrs. Creery; she
never did that ! and peeping over the blind,
she beheld to her amazement, Mr. Latimer
and Mrs. Home.   And Mrs. Home was
crying, what could it be ?   And they were
both coming to her !

A pang of apprehension seemed to seize
her heart with a clutch of ice, some unknown, some dreadful trouble was on its
way to *her*.   She sprang down the steps and
met them, saying,—

"What is the matter ?   Oh ! Mr. Latimer,
you have come to tell me something—something," growing very white, "about papa ?"

Mr. Latimer himself was deadly pale,
and seemed to find considerable difficulty in
speaking.   At last he said,—

"Yes ; he has been hurt on the ranges."

"Then let me go to him at once—at once."

"Oh, my dear, my dear," cried Mrs.

Home, bursting into tears, "you must pre-
pare yourself for trouble."

"I am prepared; please let me go to him.
Oh! I am losing time; where is he? Why,
they are bringing him home!" as her quick
ear caught the heavy tramp of measured
feet, bearing some burden—an hospital
dhoolie.

Before either of her visitors had guessed
at her intention, she had flown down the pathway, and met the procession. She hastily
pulled aside the curtain, and took her father's
hand in hers. But what was this? this motionless form, with closed eyes? She had never
seen it before in all her life, but who does not
recognize Death, even at their first meeting?

"Oh! he is dead," she shrieked, and fell
insensible on the pathway.

For a long time she remained unconscious,
and "it was best so" people whispered.
There were so many sad arrangements to be
made. The General himself superintending
everything with regard to the funeral, which
was to take place at sundown, as was the in-
variable custom in the East. There, there

is no gradual parting as in England, where white-covered dead lies amid the living for days. In India such hospitality is never shown to death, he is thrust forth the very day he comes. The wrench is agonizing, and, as in a case like the present, where death was sudden, the shock overwhelming.

To think that you may be laughing and talking with a relative, friend, or neighbour, one evening, that they have been in the very best of health, as little anticipating the one great change as yourself, and that by the very next night they may be dead, and *buried!* In Eastern countries, there seems to be almost a cruel promptness about the funerals, but it is inevitable. By five o'clock, everything was ready in the bungalow on the hill; the bier and bearers, the mourners, the wreaths of flowers, and the Union Jack for pall. Colonel Denis had that morning, been given a huge bunch of white flowers for Helen; lovely lilies, ferns, and orchids, that did not grow on Ross; he had brought home and presented the offering with pride, and she, being unusually lazy, had left the flowers

in a big china bowl, intending to arrange
them after breakfast.

How little are we able to see into the
future! Happily, for ourselves. — Would
Colonel Denis have carried home that big
bunch of lilies, with such alacrity, had he
known that they were destined to decorate
his own coffin!

In deference to Helen, who was now alive
to every sound, the large *cortège* almost
stole from the door, and the band was mute.
The cemetery was on Aberdeen, not far from
the fatal ranges, and the funeral went by
boat. Once on the sea, that profoundly
melancholy strain, "The Dead March in
Saul," was heard, after three preliminary
muffled beats of the drum; and it sounded,
if possible, more weird and sad than usual.
As its strains were wafted across the water,
and reached the bungalow on the hill, Helen
sat up on the sofa, and looked wildly at Mrs.
Home and Mrs. Durand.

"I—I—hear—the 'Dead March' in the
distance! Who—who is it for? It is not
playing for papa.—It is impossible, *im-*

possible. See, here are some of the flowers he brought me this morning—there are his gloves, that he left to have mended! I know," wringing her hands as she spoke, " that people do die, but never—never like this! This is some fearful dream; or I am going mad; or I have had a long illness, and I have been off my head. Oh, that band—" now putting her fingers in her ears, and burying her face in the cushions, "it is a dream-band—a nightmare!"

After a very long silence, there was another sound from across the water—the distant rattle of musketry repeated thrice, and now Mrs. Home and Mrs. Durand were aware that the last honours had been paid to Colonel Denis,—who had been alive and as well as they were that very morning,—and was now both dead and buried.

·   ·   ·   ·   ·   ·   ·

Nothing short of the very *plainest* speaking had been able to keep Mrs. Creery from forcing herself into Helen's presence. But Mrs. Home, Mr. Latimer, and Dr. Malone, were as the three hundred heroic Greeks, who kept the pass at Thermopylæ. They formed a body-guard she could not pass.

Every one, even the last-mentioned matron, desired to have Helen under their roof. Mrs. King came up from Viper, all the way in the midday sun, to say that, " Of course, every one *must* see, that the further Miss Denis was from old associations, the better, and that her room was ready." Mrs. Grahame arrived from Chatham, with the same story; but in the end, Helen went to Mrs. Home, going across with her after dark, like a girl walking in a trance. Sleep, kind sleep, did come to her, thanks to a strong opiate, and thus, for a time, she and her new acquaintance, grief, were parted. The pretty bungalow on the side of the hill, so bright and full of life only last night, was dark and silent now. One inmate slept a sleep to deaden sorrow, the other lay alone upon the distant mainland, under the silent stars, within sound of the sea—and the new cemetery contained its first grave.

CHAPTER X.

"Joy comes and goes, hope ebbs and flows,
                    Like the wave.
Change doth unknit the tranquil strength of man ;
        Love lends life a little grace,
        A few sad smiles ; and then,
        Both are laid in one cold place,
                    In the grave."

M. Arnold.

Days crawled by, and Helen gradually and painfully began to realize her lot. Hers was a silent, stony grief (now that the first torrent of tears had been shed) of that undemonstrative, reserved nature, that it is so difficult to alleviate, and that shrinks from outward sympathy. People, (ladies,) came to her, and sat with her, and held her hand, and wept, but she did not; this grief that had come upon her unawares, seemed almost to have turned her to stone. She opened her heart to Mrs. Home only ; and in answer to affectionate attempts at consolation, she said,—

"I sometimes sit and wonder, wonder if it is *true!* You see, Mrs. Home, my case is so different to others. Now, if you were to lose one child—which heaven forbid! —you have still eight remaining; if Colonel Home was taken from you, you have your children; but *I* have no one left. Papa was all I had, and I am alone in the world; I can scarcely believe it!"

"My dear, you must not say so! You have many friends, and friends are sometimes far better than one's own kin. Then there is your aunt. I wrote to her myself last mail."

"Aunt Julia! She is worse than nobody. She is an utter stranger, in reality, a complete woman of the world. She and I never got on; she was always saying hard things about *him!*"

"Well, you won't be with her long, you know! and you cannot say that you are alone in the world; you know very well that you will not be alone for long, you understand,"

squeezing her fingers significantly as she
spoke.

Helen did understand, and coloured vividly.
It seemed to her almost a sin to think of Gilbert
Lisle now, when every thought was dedicated
to her father, when all ideas of love or a lover
had been, as it were, swept out of her mind
by the blast of her recent and terrible cala-
mity.

Mrs. Home noticed the blush, but again
attributed its cause to the wrong person.

.   .   .   .

Colonel Denis' effects were sold off in the
usual manner; his furniture, boat, and guns,
were disposed of, his servants dismissed, and
his papers examined.   And what discoveries
were not made in that battered old despatch-
box! Not of money owing, or startling
unpaid bills, but of large sums due to him;
borrowed and forgotten by impecunious acquaintances—one thousand rupees here, three
thousand rupees there, merely acknowledged
by careless, long-forgotten I. O. U.'s.   Then
there were receipts for money paid,—drained
away yearly by his father's and wife's

creditors—his very pension was mortgaged. How little he appeared to have spent upon himself. All his life long he had been toiling hard for other people, who gaily squandered in a week, what he had accumulated in a year ; a thankless task ! a leaden burden !

Apparently, he had begun to save of late, presumably for Helen ; but, including the auction, all that could be placed to his daughter's credit in the bank, was only four hundred odd pounds !

" Say fifteen pounds a year," said Colonel Home, looking blankly at Mr. Creery.

" I know he intended to insure his life, he told me so last week."

" Ah ! if he only had. What is to become of the poor girl ? " continued Colonel Home ; " fifteen pounds a year won't even keep her in clothes, let alone in food and house-room. I believe he had very few relations in England, and see how some of his friends out here have fleeced him ! "

" They ought to be made pay up," returned Mr. Creery. " I'll see to *that*," he added with stern, determined face.

"How can they pay up? The fellows who signed those," touching some I. O. U.'s, "are dead. Here's another, for whom Denis backed a bill; he went off to Australia years ago. I wonder Tom Denis had not a worse opinion of his fellow-creatures."

"In many ways, Tom was a fool; his heart was too soft, his eyes were always blind to his own interests : some people soon found that out."

"Well! what is to become of his daughter? That is what puzzles me," said his listener anxiously. "She is a good girl, and uncommonly pretty!"

"Yes; her face is her fortune, and I hope it will stand to her," rejoined Mr. Creery, dubiously. "But, to set herself off, she should go into fine society and wear fine clothes, and she has no means to start her in company where she would meet a likely match. As they say in my country, 'Ye canna whistle without an upper lip.'"

"She might not have *far* to go for a husband," returned Colonel Home significantly.

"Ah, well! I believe I *know* what you

mean, but that man will be needing a fortune.
He is too cannie to marry 'a penniless lass
without a lang pedigree!'"

"My wife has her fancies," said Colonel
Home, "and thinks a good deal of him."

"So does mine," returned the other, "and
has *her* fancies too; but all the same—
between you and me, Home—I never liked
the fellow; you know who I mean. He is
just a gay popinjay, taking his turn out of
everybody that comes in his way."

(Observe, cannie Scotchman as he was,
that all this time, he had never mentioned
any *name*.)

.   .   .   .

Several doors were opened to Helen, offering her a home, but she steadily resisted all
invitations. She felt that she would be occupying an anomalous position by remaining on
at Port Blair, without having any real claim on
any one in the settlement. If there had been
some small children to teach,—save those in
the native school,—or if there were any means
by which she could have earned her liveli-
hood, it would have been different; but, of

course, in a place like the Andamans, there was no such opening. The community were extremely anxious to keep her among them, and were kinder to her than words could express. Mrs. Grahame besought her most earnestly to remain with her as a sister, and urged her petition repeatedly.

"The favour will be conferred by *you*, my dear, and you know it," she said. "Think of the long, lonely days I spend at Chatham, cut off from all society in bad weather, and in the monsoon, I sometimes don't see another white woman for weeks. Imagine the boon your company would be to me. Remember that your father was an old friend of Dick's, and say that you will try us for at least a year. We will do our very best to make you happy."

And other suggestions were delicately placed before Helen. Would she remain, not as Miss Denis, but as *Mrs.* Somebody? To one and all, she made the same reply, she must go home, at least, she must go back to England; her aunt had written, and desired her to return at the first opportunity, and her aunt was her nearest relation now, and all her future plans were in her hands. Mrs. Home was returning in March, they would sail together.

"If I were not obliged to place Tom and Billy at school, and see after my big boys, I would not *allow* you to leave at all, Helen," said her friend and hostess decidedly, "but would insist on your remaining with us as one of our family, a kind of eldest daughter."

Nevertheless, Mrs. Home cherished strong but secret hopes that her young *protégée* would stay at Port Blair, in spite of her own departure. Was not Mr. Quentin expected from Camorta by the very next mail?

Mrs. Creery would have liked Helen to remain with some one (not herself, for she was not given to hospitality). She considered that she would be a serious loss to the community, and was quite fond of her in her own way. Why should she not marry Jim Quentin? was a question she often asked herself in idle, empty moments. It would be a grand match for a penniless girl; a wedding would be a pleasant novelty, no matter

how quiet, and she herself was prepared to
give the affair her countenance, and to
endow the young couple with a set of plated
nut-crackers that had scarcely ever been
used! One day, roaming rather aimlessly
through the bazaar, she came across "Ibrahim," Mr. Quentin's butler, and was not the
woman to lose a rich opportunity of cross-
examining such an important functionary.
She beckoned him aside with an imperious
wave of the hand, and commenced the conversation by asking a very foolish question,
"When did you hear from your master?"
seeing that there had been no mail in, since she
had seen Ibrahim last, "when is he expected?"

"Mr. Quentin not my master any more,"
he returned, with dignity, "I take leave that
time sahib going Nicobars."

"Having made your fortune?" drawing
down the corner of her mouth as she spoke.

"I plenty poor man, where fortune getting?" he replied, with an air of surprised
and injured innocence.

"Stuff and nonsense! you know you
butlers, make heaps out of bachelors like

Mr. Quentin, who never look at their accounts, but just pay down piles of rupees,
like the idiots they are; and what about Mr.
Lisle?"

Ibrahim grinned and displayed an ample
row of ivory teeth.

"Ah," with animation, "that very good
gentleman, never making no bobbery!
Plenty money got!"

"Plenty money! How do you know?"

"First time coming paying half—after two
weeks paying *all;*" in answer to the lady's
gesture of astonishment. "Truth I telling!
wages, boats, bazaar, and *all!*"

"And what did Mr. Quentin say?"

"Oh, laughing, telling Lisle, sahib
plenty rupees got, I poor devil! Mr. Quentin very funny gentleman, making too much
bobbery, swearing too much, throwing boots
and bottles, no money giving; I plenty
fraiding, and so I taking leave," concluded
Ibrahim majestically.

This little side-light on Mr. Quentin's
manners was a revelation to Mrs. Creery.
And so Lisle was *really* rich! the dinner she

had graced at Aberdeen (on a mutton day),
had been given at *his* expense, and all
the establishment of servants, coolies, and
boatmen had been maintained by him. She
pondered much over this discovery—and,
marvellous to relate, kept it to herself.

Colonel Denis had now been dead about
two months, and his daughter was once
more to be seen out of doors, and walking
about the island; but how different she
looked, what a change a few weeks had made
in her appearance. She was clad in a plain
black dress, her eyes were dim and sunken,
her face was thin and haggard, her figure
had lost its nice rounded outlines. She was
trying to accustom herself to her new lot in
life; to that empty bungalow on the hill-side,
that she never passed without a shudder,
for did it not represent the wreck of her
home.

Something else had also been scattered to
the winds, blown away into space like gossa-
mer-web in a gale, I mean that airy fabric
known as " Love's Young Dream."

She had been dwelling on four words,

more than she herself imagined; on the promise, " I shall come back," breathed under
the palm-trees that night, that saw " flying
between the cold moon and the earth, Cupid
all armed! "

. . . .

Helen occasionally spent a day with Mrs.
Grahame or Mrs. Durand; they liked to
have her with them, and endeavoured by
every means in their power, to distract her
mind from dwelling, as it did incessantly, on
her recent loss. One morning, as she sat
working in Mrs. Durand's cool, shady drawing-room, doing her best to seem interested
in her hostess' remarks, they heard some one
coming rapidly up the walk, and Captain
Durand sprang up the steps, and entered,
holding a bundle of letters in his hand.

" The mail is in from Rangoon," he said;
" Rangoon and the Nicobars."

If he and his wife had not been wholly
engrossed in sorting their correspondence,
they would doubtless have noticed, that their
young lady guest had suddenly become very
red, and then very white, but they were

examining their letters, with the gusto of people to whom such things are both precious and rare.

" By the way," exclaimed Captain Durand, looking up at last, " Quentin is back; I met him on the pier."

Helen almost held her breath, her heart stood still, whilst her hostess put into words a question she could not have articulated to save her life.

" And Gilbert Lisle, did you see him ? "

" Oh, no! he has gone on to Japan," responded her husband, as he carelessly tore open a note. " He is a regular bird of passage ! "

" Ah, I *thought* we should not see him again," rejoined Mrs. Durand, with a tinge of regret in her voice.

Helen listened as if she were listening to something about a stranger, she bent her eyes steadily on her work, and endeavoured to compose her trembling lips. Mrs. Durand, happening to glance at her, as, opening an envelope, she said, " Why, here's a note from him ! " was struck by the strange, dead pallor of her face, and by the look of almost desperate expectation in her eyes—eyes now raised, and bent greedily on the letter in her own hand. This change of colour, this eager look, was a complete revelation to that lady, who paused, drew in her breath, and asked herself with a thrill of apprehension, " Could it be possible that Helen had lost her heart to Gilbert Lisle ? Was *she* the attraction that had held him so fast at Port Blair ? "

As she stared in a dazed, stupid sort of way, her young friend dropped her eyes, bent her head, and resumed her work with feverish industry ; but, in truth, her shaking fingers were pricking themselves with the needle, instead of putting in a single stitch !

" A note from Lisle ? And pray what has he to say ? " inquired Captain Durand, ignorant of this by-play. " Here," holding out his hand, " give it to me, and I'll read it."

" Camorta, March 2nd.

" DEAR MRS. DURAND,—As I have changed my plans, and am not returning to Port Blair, I send you a line to bid you good-bye, and

to beg you to be good enough to accept
my small sailing-boat which lies over at
Aberdeen. You will find her much more
handy for getting about in, than the detach-
ment gig. My nets and fishing-gear I
bequeath to Durand. I am going on to
Japan, *viâ* Rangoon and Singapore, and
shall make my way home by San Francisco.
Hoping that we shall meet in England ere
long, and with kind regards to all friends at
Ross,

"I remain,

" Yours sincerely,

" GILBERT LISLE."

"By Jove!" exclaimed Captain Durand,
" that smart cutter of his is the very
thing for you, Em, and the fishing-tackle
will suit me down to the ground. I like
Lisle uncommonly, but," grinning significantly as he spoke, " this note of his, consoles
me wonderfully for his departure."

Yes, so it might—but who was to console
Helen ? She felt like some drowning wretch,
from whom their only plank has just been
torn, or as shipwrecked sailor, who had pain-

fully clambered out of reach of the waves and been once more cruelly tossed back among them.

It was only now at this moment of piercing anguish that she thoroughly realized how much she had been clinging to Gilbert Lisle's promise, how steadfastly she had believed in his words, " I shall come back."

With a feeling of utter desolation in her heart, with her ideal and her hopes alike shattered, what a task was hers to maintain an outward appearance of indifference and composure!

After a time Captain Durand went off to the mess, to hear the news, and to look over the papers, leaving the two ladies *tête-à-tête ;* his wife affected to peruse her letters, reading such little scraps of them aloud from time to time as she thought might amuse her companion, but she was not enjoying them as usual. That look she had surprised in the girl's eyes, haunted her painfully. She longed to go over to her, and put her arm round her neck and whisper in her ear,—

"What is it? Tell me all about it, confide in *me*."

But somehow she dared not, bold as she was.—Recent grief had aged Helen, and given her a gravity far beyond her years, and as she looked across at that marble face, those downcast eyes, and busy fingers, she found her kind, warm heart fail her. Whatever the hurt was, ay, were it mortal, that girl meant to bear it alone.

She was more affectionate and sympathetic to her young friend than usual, smoothed her hot forehead, kissed her, caressed her, and whilst they sat together in the twilight in the verandah, looking out on the dusky sky, found courage to murmur,—

"Dearest Helen, remember that I am your friend, not merely in name only. Should you ever have any—any little trouble such as girls have sometimes, you will come and share it with me, won't you? I am older, more experienced by years and years, and I will always keep your secrets, exactly as if they were my own!"

This was undoubtedly a strong hint; nevertheless, her listener merely smiled and nodded her head, but made no other sign. "*Little trouble!*" She was on the rack all day long. She bore the torture of her hostess's soft whispers and tender, sympathetic looks, which told her that she guessed *all*. She bore the brightly-lit dinner-table, and Captain Durand's cheerful recounting of the most thrilling news. She even endured his eloquent praises of Gilbert Lisle without flinching. Little did her gallant host guess the effort that those smiles and answers cost her. Good, commonplace man! he had got over his brief love affair fifteen years previously, and had forgotten it as completely as a tale that is told. Mrs. Durand had a more vivid recollection of her own experiences,—and a share of that fellow-feeling that makes us all akin. She was amazed at Helen's fortitude, especially when she glanced back over the past and remembered (and I hope this will not be put down to her discredit) that when *she* had seen the announcement of the marriage of her first fancy in the paper, she had spent the re-

M 2

mainder of the day in hysterics and the subsequent week in tears. She walked back with Helen, and left her herself at Colonel Home's door, and bade her good-night with unusual tenderness. Then she retraced her steps, arm in arm with her husband, whose mind was abruptly recalled from planning a long day's sea-fishing, by her saying rather suddenly,—

"I know *now* why Helen refused Dr. Parks!"

"Oh!" contemptuously, "I could have told you the reason long ago, if you had asked me. Because he was the same age as her father!"

"No, you dear, stupid man—but this is quite private. I am sure," lowering her voice, "that she likes Gilbert Lisle."

A long whistle was the only reply to his information for some seconds, and then he said,—

"Now what has put *that* into your head?"

"Her face, when you came in and told us that he was not coming back. I cannot get it out of my mind, it was only a momentary expression, she rallied again at once ; but that moment told me a tale that she has hitherto guarded as a secret."

" You are as full of fancies and ridiculous, romantic ideas as if you were seventeen instead of—"

" Don't name it ! " she interrupted hastily, " the very leaves here have ears ! "

Her husband laughed explosively, and presently said,—

" I never knew such a woman as you are for jumping at conclusions. She had a twinge of face-ache, that was all."

" A twinge of heart-ache, you mean. But what is the use of talking to *you* ?—you are as matter-of-fact as a Monday morning. And now, pray tell me, though I suppose I might just as well ask Billy Home, did Gilbert Lisle ever show her any attention ? "

" Ha—hum—well, do you think that saving her life could be called an attention ? "

" Yes," eagerly ; " yes, of course ! I'd forgotten about that ! "

" And another time he picked her off the mainland and brought her home in what is now your boat, through a series of white squalls."

" Did he really ? " the really, as it were, in large capitals.

" And he was there a few times. But you need not get any ideas into your head about *him*, it was always Quentin, he was always hanging about her in that heavy persistent way of his—it was Quentin, I tell you ! "

" And *I* tell you," responded his wife emphatically, " that it was, and is, Gilbert Lisle. I recollect his saying, the night of the ball, what a nice girl she was ; or *I* said it, and he agreed, which is the same thing. And I remember perfectly, now that I think of it, noticing them leaning over a gate, and looking just like a pair of lovers."

A loud and rudely incredulous haw-haw from Captain Durand was his only reply.

" You may laugh as much as you like, but Mr. Lisle told me that he would gladly give a thousand pounds to get out of the Nicobars trip, and the last thing he said to me, as he bade me good-bye, was, ' I shall see you again soon.' I remember all these things now, and put two and two together, but I cannot make it out—I am utterly puzzled. Perhaps Mr. Quentin will be able to throw some light on the subject!"

"Quentin wants to marry her himself."

"Not he! He only wished to be a dog in the manger, to engross the only pretty girl in the place, that was all. I know him *well*. And now that she has been left an orphan, without a fraction, he has as much idea of making her Mrs. Quentin, as he has of flying over the moon!"

"All right, Em, time will tell.—I bet you a new bonnet that this time next year, she will be Mrs. Q."

"No more than she will be Queen of England," returned his wife with emphasis. This was positively the last word, and Mrs. Durand's property, for they had now reached the steps of their own bungalow, and consequently the end of their journey.

CHAPTER XI.

" FAREWELL, PORT BLAIR."

" Farewell at once—for once, for all—and ever."
Richard II.

Mrs. Durand's surmises were correct.

A few days after James Quentin's return, without any marked haste, he went over and called on Mrs. Home and Miss Denis. The former was an arrant little match-maker, and was delighted to see that *débonnaire* face once more. He was handsome, rich (?), and agreeable, he had been devoted to her young friend previous to his departure for the Nicobars, and, *of course*, it would be all settled now. With this idea in her head, she presently effaced herself so as to give the gentleman ample opportunity for a *tête-à-tête.* She even kept Tom and Billy out of the way, and this was no mean feat.

Mr. Quentin murmured some polite stereotyped regrets, then he alluded in rather strong language to "that vile hole Camorta." As he talked he stared, stared hard at Helen, and wondered at the change he saw in her appearance. She was haggard and thin; of her lovely colour not a vestige remained, and the outlines of her face were sharp, and had lost their pretty contour. She looked like a flower that had been beaten down by the storm. Never in all his experience had he beheld such a complete and sudden alteration in any one; he was glad he had never thought of her seriously, and as to Lisle, he was well out of it (thanks to his friend James Quentin); *he* took everything so seriously he would have been sure to have got the halter over his head, and to have blundered into an imprudent match. His yes meant yes; his no, no. Now he himself had a lightness of method, a nebulous vagueness surrounded his most tender speeches; at a moment's notice, he could slip off his chains, and run his head out of the noose, and always without any outward unpleasantness— that was the best of the affair. Gilbert Lisle was different, he was not used to playing with such brittle toys as girls' hearts. Well, this girl had entirely lost her beauty, so thought her visitor, as he contemplated her critically and conversed of malaria and Malays. She had not a penny, and no connections; he supposed, when she went back to England, she would go out as a governess, or a companion, or music-teacher. He entirely approved of young women being independent and earning their own bread. If there was a subscription got up for her passage money, he meant to do the handsome thing, and to give fifty rupees (5*l.*).

"I suppose you were surprised to hear about Lisle?" he said at last.

"Yes," looking at her questioner with complete composure.

"He left me at Camorta, you know. He is a queer, eccentric beggar, and you would never suppose, to see him in his old fishing-kit, and with his hands as brown and horny as a common boatman's, that he had been in the Coldstreams, and was a regular London swell."

Helen made no reply, and he continued glibly,—

"He is considered a tremendous catch; they say his elder brother is dying at Algiers—consumption—but he is not easy to please!"

"Is he not?" she echoed with studied indifference.

"No.—By Jove! Mrs. Creery did not think much of him; she was awfully rough on him. How all you people did snub him! Many a good laugh I had in my sleeve!" and he smiled at the recollection.

"I do not think that many people snubbed him," returned Helen with a flushed cheek and flashing eye.

"Well, perhaps *you* did not," returned Mr. Quentin, somewhat abashed. "You know, you never snubbed any one but me," with a mental note that she should live to be sorry for that same. "Lisle made me promise to keep his secret. He wished to be accepted for himself for once, without any *arrière pensée* of money or title; and by George, he got what he wanted with a vengeance—eh? I don't think he will try it again in a hurry.

He found his level,—the very bottom of the ladder, something quite new!" and again he laughed heartily at the recollection.

"I suppose it was," with elaborate indifference.

"He had been having a big shoot in the Terai before he came here. He was awfully taken with this place, the queer, unconventional life, and stayed on and on, greatly to my surprise. Many a time I wondered what he saw in the place, though, of course, I was delighted to have him. My luck was dead in." (So it was, *vide* Ibrahim's domestic accounts!)

"Yes, of course it was pleasant for *you*," admitted Helen.

"He should have been a poor man; he had so much energy and resource, and such Spartan tastes. Ten times a day I wished that we could change places."

"I daresay!" returned the young lady rather drily.

There was something—was it a tone of lurking scorn?—in this "I daresay!" that irritated her listener, who instantly resolved to administer a rap on the knuckles in return.

"His father is wild with him for roving about the world; he wants him to marry and settle."

" Yes ? "

"I believe he has an heiress in cotton-wool for him at home.—I wish my governor was as thoughtful ! "

"No doubt he knows that *you* are quite equal to finding such a treasure for yourself," returned Miss Denis, with a very perceptible touch of sarcasm.

Mr. Quentin laughed rather boisterously. It was new to him to hear sharp speeches from ladies' lips, and now, looking at his watch, and rising with a sudden start, he said,—

"I declare ! I must be going ! I had no idea it was so late. I've an appointment (imaginary) at four o'clock, and I've only two minutes. Well," now taking her hand, "and so you are off on Wednesday ? I may see you before that, if not, good-bye," holding her fingers with a lingering pressure, and looking down into her eyes as if he felt unutterable regret, quite beyond the reach of words; but in truth he was conscious of nothing, beyond a keen desire to make a happy exit, and to get away respectably (perhaps he had also a lurking craving for a "peg!"). "Good-bye, I hope we shall meet again some day in England. Perhaps you would drop me a line?" a query he had often found to have an excellent and soothing effect at similar partings.

Helen took no notice of the suggestion, but merely bowed her head and said very quietly,—

"Good-bye, Mr. Quentin, good-bye."

And then the gentleman took himself away in exaggerated haste, muttering as he hurried down to the pier,—

"How white she looked, and how stiff she was! I'm hanged if I don't believe she had a weakness for Lisle, after all. If *that's* the case, this humble, insignificant, individual, has put a pretty big spoke in her wheel."

.   .   .   .

It is almost needless to mention that Helen

was now accustomed to daily interviews with
Mrs. Creery, and to being cross-examined as
to how she had been left, whether Mr. Quentin had said " anything," and what she
" was going to do with all her coloured
dresses ? "

Eliza Creery was a pertinacious woman,
and had not lost sight of her designs upon
the black silk gown (neither had Helen).

" My dear," she said, " if you ask my
advice," the last thing that was likely to
occur to her listener, " you will sell all your
things. They will be a perfect boon here,
and it is not unusual in cases of sudden
mourning, and utter destitution, such as
yours ! " Helen winced and grew very pale.
" I really think that you might have had this
made with a little more style," touching her
black dress. " But now," seriously, " *what*
about your others ? "

" Lizzie Caggett was asking about my
cottons."

" Yes ? " stiffening with apprehension.

" I told her that I would be only too glad
to let her have them ! There are one or two

that I cannot bear to look at. *He* liked them," she added under her breath.

" And for how much ? What did you ask for them ? "

" Why, nothing, of course ! " returned Helen in amazement.

" Then she shan't have them ! I shall not stand by and see you fleeced ! I shall certainly speak to her mother ! What a horrible, grasping, greedy girl ! Taking advantage of your innocence—she would not get round *me* like that ! " (Mrs. Creery never spoke a truer word).

" But they are useless, quite useless to me," exclaimed Helen.

" Rubbish ! nonsense ! is *money* useless to any one ? Did you give her anything else ? " demanded the matron sharply.

" Only my best hat, and a few new pairs of *gants de Suéde*."

" This must be stopped *at once:* She has no conscience, no principle.—You will be giving her your white silk next, you foolish girl. You must think of yourself, you have hardly a penny to live on, and are as lavish as a princess, and utterly indifferent to your own interests. Now, if you had spoken to *me*, I could have disposed of your cottons and muslins for ready money. As it is, I shall take your black silk, your white silk, your blue surah," running over these items with infinite unction, " and give you a good price for them, considering that they are second-hand. Your white satin low body would be too small, I'm afraid; and your gloves are not my size (Mrs. Creery took sevens, and Helen sixes) ; but I'll have your pinafores and brown hat."

" But, indeed, thanking you very much for thinking of me, I do not wish to sell anything. Some day I may want these things, and have no money to replace them, don't you see ? "

Mrs. Creery failed to see the matter in that light at all, and argued and stormed ; nevertheless, Helen was adamant.

" Aunt Julia would not be pleased, I'm sure," she said firmly. " And I really could not do it, really I would not, Mrs. Creery."

" And I had such a fancy for your little

black lace and jet shoulder-cape!" whimpered
that lady, on the verge of tears.

Helen paused, looked at her hesitatingly,
and said,—

"I wonder if you would be very much
offended if—if I—" here she broke down.

But Mrs. Creery knew exactly what she
wished to say, and rushed to her rescue.

"Yes, that's it exactly," she cried eagerly,
"a *capital* idea, we will exchange! I'll
take your cape, which would be brown next
year, and give you something you will like
far better, something that won't wear out,
and will serve to remind you of the six
months you spent at Port Blair." (As if
Helen needed anything to remind her of
that.) "Something that, I'm sure, you will
be delighted to have."

On these conditions the barter was agreed
to, and the elder lady folded up and carried
away the cape. Doubtless she feared that
Miss Denis might yet change her mind!

The same afternoon Mrs. Creery's ayah
sauntered down with a small paper parcel in
her hand, and when it was opened, Helen

discovered an exceedingly trumpery pair of shell bracelets, tied with grass-green ribbon —total value of these ornaments, one Government rupee, in other words, eighteen pence !

Mrs. Home, who had heard of the fate of the little shoulder-cape, became quite red with indignation, and was loud (for her) in her denunciation of Mrs. Creery's meanness. But Helen was no party to her anger and scorn, nay, for the first time for many weeks, she laughed as merrily and as heartily as she had been wont to do in the days that were no more.

.   .   .   .

The eventful Wednesday came that brought the English letters, and took away Mrs. Home, and Helen.—The whole community rowed out to the *Scotia* to see them off, laden with books, and flowers, and eau-de-Cologne, and fruit. When I say the whole community, Mr. Quentin was the exception that proved the rule. Jim Quentin was conspicuous by his absence, and neither note nor bouquet arrived as his deputy. Mrs.

Home was keenly alive to his defection and extremely put out, though her anger smouldered as fire within her, and she never breathed a word to Helen, and thought that she had never seen a girl bear a disappointment so beautifully.

There was maiden dignity! There was fortitude! There was self-control! Mrs. Durand hung about her friend with little gifts and stolen caresses,—she had not failed to notice that Apollo was not among the crowd, and had whispered to her husband, as they stood together, "*He* is not here, you see, and the bonnet is *mine*."

To Helen she said,—

"Mind you write to me often; be sure you do not drift away from me, my dear. When I go home, you have promised to come and see me, and, you know, you would be going to my people now, only they are in Italy at present. Be sure you don't forget me, Helen."

"Is it likely?" she returned. "Have I so many friends? Do not be afraid that I shall not write to you often, perhaps *too* often. I shall look out for your letters far more anxiously than you will for mine, and is it likely that I can ever forget you? You know I never could."

Mrs. Creery was present of course, and when time was up, and the bell rang for visitors to descend to their boats, she actually secured the last embrace, saying as she kissed Helen on either cheek,—

"So sorry you are going, dear. Of course, you will write? I have your address: 15, Upper Cream Street. It has all been very sad for you, but life is uncertain;" then as a *bonne bouche* reserved for the last, a kind of stimulant for the voyage,—she added impressively, "My sister, Lady Grubb, will call on you in London—and now, really, good-bye." One more final whisper yet in her ear, positively the last word, "Quentin has treated you disgracefully."

A pressure of the hand and she was gone.

The steamer's paddles began to churn, to grind the water, the boats rowed on alongside, their occupants waving handkerchiefs, till the *Scotia* gradually forged ahead and left them all behind.

Helen leant over the bulwarks, watching them, and waving to the last. How much she liked them all, how good they had been to her! As they gradually fell far behind, even the final view of Mrs. Creery's broad back and mushroom topee caused her a pang of unexpected regret.

The surrounding hills, woods, and water looked lovelier than she had ever seen them, as if they were saying, " How can you bid us ·good-bye ? Why do you leave us ? "

She gazed with straining vision towards the graveyard on the hill, now fading so fast from eyes that would never see it more. Presently Mount Harriet became sensibly diminished, then Ross itself dwindled to a mere shadowy speck; Helen stood alone at the taffrail, taking an eternal farewell of these sunny islands, which had once been to her as an earthly paradise, where the happiest hours of her life had been spent, and the darkest—where she had first made acquaintance with love and death and grief! The

little-known Andamans were gradually fading
—fading—fading. As she stood with her
eyes earnestly fixed upon the last faint blue
outline, they were gone,—merged in the
horizon, and lost to sight. She would never
more behold them, save in her dreams!

With this thought painfully before her
mind, she turned slowly away, and went
below to her own cabin, and shutting fast
the door, she threw herself down on her
berth, and wept bitterly.

CHAPTER XII.

THE STEERAGE PASSENGER.

"Pray you sit by us, and tell's a tale."
Twelfth Night.

"MRS. HOME and party" were to be seen in the list of names of those who sailed from Calcutta in the steamer *Palestine* on the 20th of March. There were not many other passengers, but those on board were sociable and friendly; and the old days, when Bengal and Madras did not speak, paraded different sides of the deck, and only met in the saloon at the point of the knife (and fork), are gone, to return no more! The weather was at first exceedingly rough, the water "plenty jumping," in the phraseology of Mrs. Home's ayah. She, like her mistress, became a captive to Neptune almost as soon as the engines were in motion. Once out on the open sea she lay for days on the floor, rolled up in her sarée like a bolster or a mummy, uttering pitiful moans and invocations to her relations ! Helen was a capital sailor, and took entire charge of Tom and Billy, and was invaluable to her sick friend, upon whom she waited with devoted attention, tempting her with beef-tea and toast and other warranted sea-refreshments.

Not a few of her fellow-passengers would have been pleased to while away the empty hours in dalliance with the tall girl in black, but she showed no desire for society, and as it was whispered that she had recently lost some near relation, and was *really* in deep grief, she was left to herself, and to the company of Tom and Billy.—It seemed quite marvellous to the community, that such a pretty girl should be returning to England *unmarried.* They shrugged their shoulders, lifted their eyebrows, and wondered to one another whether it was because *she* was too hard to please, or whether the community at Port Blair were stolid semi-savages ?

The first little piece of excitement that broke the monotony of the voyage, was the discovery of a stowaway in one of the boats, who was not starved out till they had passed Galle. He proved to be a deserter from a regiment in Calcutta, and was promptly sent below to stoke, as extra fireman, and doubtless he found that employment (especially in the Red Sea) even less to his taste than drilling in the cool of the morning on the Midan near Fort William. The Red Sea was as calm as the proverbial mill-pond, and the motion of the steamer almost imperceptible. The ayah recovered from her state of torpor, and Mrs. Home actually made her appearance at meals, and joined the social circle on deck. Every evening there was singing, the songs being chiefly contributed by the ladies and one or two German gentlemen *en route* from Burmah to the Fatherland. Passengers who could not, or would not, perform vocally, were called upon to tell stories, and those hot April nights, as they throbbed past the dark Arabian coast, were long remembered by many on board. Chief among the entertainers was the captain of the *Palestine*. He related more than one

yarn of thrilling adventures by sea. The
German merchants told weird legends, and
episodes of the late greàt war, a grizzled
colonel gave his experiences of the Mutiny,
a subaltern his first exploit out after tigers,
but the most popular *raconteur* of them all
was the first officer, Mr. Waters. When he
appeared, and took his seat among the company after tea, there was an immediate and
clamorous call for a story—a story.

"Now, Mr. Waters, we have been waiting
for you!"

Apropos of the stowaway, he recounted the
following tale, to which Billy Home, who was
seated on Helen's knee, with his arm encir-
cling her neck, listened with very mixed sen-
sations :—

"When I was second officer of the *Black
Swan*, from Melbourne to London," he began
promptly—yes, he liked telling yarns,—"we
had one uncommonly queer trip, a trip that
I shall not forget in a hurry, no, and I don't
fancy that many of those who were on board
will forget it either! It was the year of the
Paris Exhibition, and all the world and his

wife were crowding home. We had every berth full, and people doubled up anywhere, even sleeping on the floor of the saloon. We left port with three hundred cabin, and seventy-five steerage passengers. At first the weather was as if it were made to order, and all went well till about the third night out, when the disturbance began, at least, it began, as far as *I* was concerned. I was knocked up about an hour after I had come off watch, and out of my first sleep, by some one thundering at my door. I, thinking it was a mistake, swore a bit, and roared out that they were to go to the third officer, and the devil ! But, instead of this, the door was gently opened, and the purser put in a very long white face, and said,—

" ' Look here, Waters, I want you in my cabin; there is the mischief to pay, and I can't make it out ! I can't get a wink of sleep, for the most awful groans you ever heard ! '

" I sat up and looked at him hard. He was always a sober man, he was sober now, and he was not walking in his sleep. After

a moment's very natural hesitation, I threw
on some clothes, and followed him to his
cabin, which was forward. The light was
still burning, and his bunk turned back just
as he had leapt out of it; but there was
nothing to be seen.

"'Wait a bit,' he said eagerly, 'hold on a
minute and listen.'

"I did, I waited, and listened with all my
ears, and I heard nothing but the thumping
of the engines, and the tramping of the
officer overhead on watch. I was about to
turn on my heel with rather an angry
remark, when he arrested me with a livid
face, and said,—

"'There it is!' and sure enough there
it *was*,—a low, deep, hollow groan, and no
mistake about it, a groan as if wrung from
some one in mortal agony, some one suffering
lingering and excruciating torture.—I looked
at the purser, big beads of perspiration were
standing on his forehead, and he looked hard
at me. 'I heard it all last night,' he said in
a husky whisper, 'but I was afraid to speak.
I hunted to-day high and low, and sounded

every hole and corner, but there is nothing to be found ! ' Then he ceased speaking, there it was *again*, louder, more painful than ever ; it certainly came from some place below the floor, and on the starboard side. We both knelt down, and hammered, and knocked, and called, and laid our ears to the boards, but it was of no use,—there was silence.

" ' Perhaps it was some one snoring,' I suggested, ' or it might be a dog ? ' "

" ' No,' returned the purser, who was still on his knees, ' it's a human voice, and the groans of a dying man, as sure as I'm a live one ! '

" I remained in the cabin for half an hour, and though we overhauled the whole concern, we heard nothing more, so I fetched up for my own bunk, and turned in and went to sleep.

" The next day the purser said he heard the moans very faintly, as if they were now getting weaker and weaker, and after this entirely ceased. For a good spell everything went along without a hitch, we had A 1 weather, and made first-class runs. But one evening,

in the twilight, I noticed a great commotion
in the saloon, I heard high talking—a
woman's voice! One of the lady passengers
was the centre of a crowd, and was making
some angry complaint to the captain.

" ' It's the young man in the boots again ! "
she said. ' And it's really too bad. Why is
he allowed in this part of the ship, what are
the stewards about ? It is insufferable to be
persecuted in this manner ! Every evening,
at this hour, he comes to the door of the
saloon and beckons to *me*, or to any one who
is near, but he never seems to catch any one's
eyes but *mine!* It's really disgraceful that
the steerage passengers should be allowed
among us in this way.'

" The saloon stewards were all called
up and rigidly cross-examined by the captain, but they all most positively declared
that no stranger had been seen by them,
nor was there any steerage passenger
on board that at all answered the lady's
description.

" ' Of course, that's nonsense ! ' she exclaimed indignantly. ' He comes to the bar

for spirits on the sly—and very sly he is—
for I've gone to the door to see what he wanted,
and he has always contrived to slip away.'

" An extra sharp look-out was accordingly
kept by the captain's orders, but the head
steward privately informed me with a grin
' that there was no such person as a tall young
man in a blue jumper, with long boots, on
the ship's books,' and we both came to the
conclusion that the lady was decidedly
wanting in her top gear.

" ' However, after a while other people
began to see the steerage passenger. Not
merely ladies only, but hard-headed, prac-
tical, elderly men ; and very disagreeable
whispers began to get afloat that the ship was
haunted ! " The apparition in long butcher
boots, could never be caught or traced, but
he was visible repeatedly ; and did not
merely confine himself to hanging about and
beckoning at the saloon door—he was now
to be met in passages, at the dark turns on
the stairs, behind the wheel-house, and even
on the bridge,—but always after dusk.
Things now began to be extremely un-

pleasant, discipline was scorned, at the very *idea* of taking away the lights at eleven o'clock, there was uproar, and an open mutiny among the ladies. Passengers were completely unmanageable, the women going about in gangs, and the very crew in couples. The captain endeavoured to make a bold stand against the ghost, but he was silenced by a clamour of voices, and by a cloud of witnesses who had all *seen* it, and, to make matters better, we came in for the most awful weather I ever experienced, our hatches were stove in, our decks swept, and I never was more thankful in all my life, than when we took up our pilot in the Downs. What between the ghost and the gales, even our most seasoned salts were shaky, and grumbled among themselves, that one would almost imagine, that we had a dead body on board! However, we managed to dock without any misadventure, beyond being five days over our time, having lost three boats, and gained the agreeable reputation of being a haunted ship! When we were getting out the cargo, and having the usual overhaul below, I happened to be on duty one day, when I was accosted by the boatswain, who came aft to where I was standing, with an uncommonly grave face. 'Please, sir,' said he, 'we've found something we did not bargain for; it was in the place where the anchor-chain is, and now, the chain being all paid out, it's empty, you see—' he paused a moment,—'all but for a dead man.'

" Of course I hurried forward at once, and looked down into a dark hole, when, by the light of a bit of candle held by one of the crew, I saw, sure enough, crushed up against one side the skeleton of a man—a skeleton, for the rats had picked his bones clean; his coat still hung on him, he wore long digger's boots, and a digger's hat covered his bare skull.

" I started back, and fell foul of the candle,—though I'm not a particularly nervous person, for I now remembered the groans I had heard in the purser's cabin.

" ' You see, sir, how it was,' said the boatswain; 'he was a stowaway, in course. When we were in dock, this place was empty. Cause why ? The anchor-chain is out, and it

seemed to this poor ignorant wretch, who
was no seaman anyway, to be just the very
spot—as it were, made for him! I've a kind
of recollection of him, too, hanging about
when we were taking in cargo. He was
young, and looked like a half-starved, broken-
down gentleman, such as you see every day
in the colony, who come out—bless their innocence!—a-thinking the nuggets is growing
on the trees, and sink down to beggary, or to
working their way home before the mast.
Ay, he thought to get a cast back,' said the
bo'sun, ' and he just walked straight into the
jaws of death. The moment we began to
weigh anchor, and the chain came reeling, and
reeling, into his hiding place, it had no
outlet but the hole at the top, and the rattle
of it and the noise of the donkey-engine
drowned his cries : he was just walled in, poor
chap, and buried up alive ! '

"Of course, we all knew, that this was
the mysterious apparition in long boots, who
had created such an unparalleled disturbance
on the passage home. Presently the remains
were decently carried away, and there was

an inquest, but nothing could be discovered about the body. We subscribed for the funeral among us, and he was buried in the nearest churchyard. We sailors are a superstitious lot, and thought we got out of it (I mean, bringing home a corpse) better than could be expected, so we gave him a respectable funeral; but there is no name on the stone cross above his head, for the only one, we knew him by, was that of the ' Steerage Passenger ! ' "

．　．　．　．　．　．　．

The chief officer brought his story to an end in the midst of a dead, nay, an awe-struck silence. People shuddered and looked nervously behind them. They were on board ship, too! Why should not the *Palestine* have a ghost of her own, as well as the *Black Swan ?*

The utter stillness was suddenly broken by a loud howl from Billy Home, who had been listening with all the power, of his unusually capacious ears,—and seemed to have but just wakened up from a sort of trance of horror. He shrieked and clung to Helen, who had whispered a hint with regard to bed-time.

"No, no, no," he would not come. "No, not alone," he added with a yell, hanging to her with the tenacity of a limpet; "not unless you stay with me.—I'm afraid of the man downstairs,—I *know* he is downstairs."

"I declare," said the bearded story-teller, "I quite forgot that little beggar was there. I never noticed him till now, or I would not have told you that yarn."

Needless to remark, his apology came rather too late. At every turn of the companion ladder, at every open door, Billy lived in whining anticipation of meeting what he called "the man in the boots," and for the remainder of the voyage he was figuratively a mill-stone round Helen's neck.

They had an uneventful passage down the Mediterranean, halting at Malta for lace, oranges, and canaries; they passed Cape Bon, then the coast of Spain, and the snow-capped Sierra Nevada. The Home boys had never beheld snow till now, and were easily induced to believe, that what they beheld was pounded sugar, and languished at the mountains with greedy eyes, as long as they remained in sight. On a certain Sunday afternoon in April the *Palestine* arrived in the Victoria Docks, London. Numerous expectant friends came swarming on board, all eagerness and expectation, but there was no one to welcome Helen,—no face among that friendly crowd was seeking hers. Being a Sunday, there was, of course, some difficulty about cabs and trains, and the docks were very remote from the fashionable quarter where her Aunt Julia resided : so she swallowed her disappointment and made excuses to herself. However, Mrs. Home, who had been met by her brother, insisted upon personally conducting her to her journey's end. First they went by rail above ground, then by rail underground, finally by cab, and after a long drive, the travellers drew up at Mrs. Platt's rather pinched-looking mansion in Upper Cream Street. A manservant answered the bell, flung wide the door with a jerk, and stood upon the threshold in dignified amazement on beholding *two* cabs, heavily laden with baggage.

Was Mrs. Platt at home ?

" No, ma'am. She and the young ladies
have gone to afternoon church; but Miss
Denis is *expected.*"

Rather a tepid reception, Mrs. Home
thought, with a secret thrill of indignation.
Much, much she wished that she could have
taken Helen with her there and then. She
hugged her vigorously, as did also Tom and
Billy; and telling her, that she would come
and see her very soon, she re-entered her cab,
and with her brother, children, and luggage,
was presently rattled away. Helen felt as
she stood on the steps, and watched those
familiar trunks, turning a corner,—that her
last link with the Andamans, and all her
recent life, was now broken.

CHAPTER XIII.

A POOR RELATION.

"Oh, she is rich in beauty, only poor!"
Romeo and Juliet.

"You had better have your big box kept in the back hall—it will scarcely be worth while to take it upstairs, and it might only rub the paper off the wall."

This was almost the first greeting that Helen received from her Aunt Julia.

"And, dear me, how thin you have grown! I would have passed you in the street," was her eldest cousin's welcome.

Mrs. Platt and her two daughters, Clara and Caroline, had returned from church, and found their expected guest awaiting them alone, in the drawing-room! "Surely one of them might have stayed at home," she said to herself with a lump in her throat and a mist before her eyes. She had latterly been made so much of at Port Blair that her present reception was indeed a bitter contrast. It undoubtedly *is* rather chilling to arrive punctually from a long journey (say, half across the world), and to find that your visit is a matter of such little moment to your relations, that they have not even thought it necessary to remain in doors to await, much less to send to meet you ! Helen felt strangely neglected and depressed, as she sat in the drawing-room in solitary state, still wearing her hat and jacket, and feeling more like a dependent, who had come to seek for a situation, than a near relation to the lady of the house. She had fully an hour in which to contemplate the situation, ere her aunt and cousins returned. They were three very tall women, and made an imposing appearance, as they filed in one after another in their best bonnets, with their prayer-books in their hands. They kissed her coolly, inquired when, and how, she had arrived, and then sat down and looked at her attentively.

Mrs. Platt was a thin, fair lady, with

handsome profile, who had married well, and
contrived to keep herself aloof from the
general wreckage, when her maiden home
was broken up; ambition was her distinctive
characteristic; she had married well, and
got up in the world, and now she hoped to
see her daughters do the same.

To effect a lodgment in an upper strata
of society, to mix with what she called the
" best people," was her idea of unalloyed
happiness.

In her grander, loftier style she was every
bit as fond of a title as our dear friend Mrs.
Creery.

Besides all this, she was a respectable
British matron, who paid her bills weekly,
went twice to church on Sunday, never darkened the door of an omnibus, or condescended
to use a postcard. Still, in her own genteel
fashion, she was a capital manager, and
generally made eighteen pence contrive to do
duty for two shillings. She was honest,
scheming, hard to every one, even to herself,
making all those with whom she came into
contact useful to her in some way; either

they were utilized as social stepping-stones,
or givers of entertainment, concert, and opera
tickets, flowers, or better still, invitations to
country houses; all her friends were expected to put their shoulder to her wheel in
some respect—either that,—or she dropped
their acquaintance. Under these circumstances it will be easily imagined, how very
unwelcome to such a lady as Mrs. Platt was
the unlooked-for return of this handsome,
penniless niece !

The Miss Platts were tall young women,
of from six, to eight and twenty years of age ;
they had unusually long necks, and carried
their noses in the air ; they were slight, and
had light eyes and eyebrows, which gave
them an indefinite, unfinished appearance;
their hair was of a dull ashen shade, and they
wore large fluffy fringes, were considered
"plain" by people who did not like them,
and "elegant-looking girls" by those who
were their friends.

They were unemotional, critical, and selfish,
firmly resolved to get the best of whatever
was going for the Miss Platts ; influenced

their mother as they pleased, and had the greatest repugnance to having their cousin Helen thus billeted upon them.

They called everything, and every person, that did not meet with their approval " bad style," and worshipped coronets, as devoutly as their parent herself.

By-and-by the new arrival had some tea, was assured that she would be " all the better for a night's rest," and was escorted to the very top of the house, by an exhausted cousin, to what her aunt called "her old room." This was true,—it was not the guest-chamber, but a very sparsely furnished apartment, on the same floor with the maids. And here her relative deposited her candlestick, nodded a condescending good night, and left her to her repose. This was her home-coming ! However, she was very tired, and soon fell asleep, and forgot her sorrows; but very early the next morning, she was awoke by the roar of the London streets, for you could call it nothing else. Mrs. Platt, though occupying a most fashionable and expensive nutshell, was close to one of the great arteries of traffic. Helen lay and listened. What a contrast to the last place where she had slept on shore, where the bugle awoke the echoes at five o'clock in the morning, where wheels and horses were absolutely unknown, and the stillness was almost solemn, only broken by the dip of an oar or the scream of a peacock! She turned her eyes to a picture pinned to the wall, facing the foot of her bed, the picture of a merry-looking milkmaid, with a pail under her arm; the milkmaid was smiling at her now, precisely as she had done less than a year ago,—when she had slept in that very room previous to starting for Port Blair. *Then* she had seemed to her imagination, to wish her good speed. Surely that gay expression seemed to augur the future smiles of fortune! Ten months ago she had stared at that picture, ere she had set out for her voyage, full of hope and happy anticipations; and now, ere the year had gone round, she was back again, her day was over, her happy home in those sunny islands among tropical seas, had vanished like a dream! She had visited, as it were, an enchanted land, where she had found father, home, friends—ay, and lover, and had returned desolate and empty-handed (save for that "sorrow's crown of sorrow"), to face the stern realities of life,—and to earn her daily bread. She gazed at the mocking milkmaid, and closed her eyes. Oh! if she could but wake and find that the last four months, had been but a horrible dream.

The Platts were late people, they scorned the typical first worm. Helen, accustomed to early (Eastern) hours, had a very long morning, entirely alone. She dared not unpack, she had no work to do, and could find no books to read; for her aunt, who was most economical in regard to things that did not make a show, did not subscribe to a library, merely took in a daily paper, and preyed on her friends, for her other literature.

Breakfast was at eleven o'clock, and during that meal letters were read, the daily programme arranged, and peoples and places discussed, whose names were totally unknown to Helen. Now and then, her cousins threw her a word or two, but there was no cordiality or friendship in their tone; it did not need that, to tell her she was not welcome, and she sat aloof in silence, feeling as if she were an utter alien, and as if her very heart was frozen. And yet these were her own flesh and blood—her father's sister and nieces—her nearest, if not her dearest! How different to Mrs. Home, Mrs. Grahame, and Mrs. Durand!—ay, even Mrs. Creery had shown her more affection, than her own aunt.

Helen soon fell into her proper niche in the family. After breakfast she went out and did all the little household messages to the tradespeople, and made herself useful, i.e. mended her aunt's gloves, and hose, wrote her notes, and copied music for her cousins.

She dined early, when her relatives lunched, as they frequently had people in the evening.

There was a kind of back room or den upon the second landing, where the Platt family sat in *déshabillé*, partook of refreshments, wrote letters, ripped old dresses, and held family conclaves. Here Helen spent most of her time, and being very clever with her needle, did many " odd jobs " for her relatives. Better this, than sitting with idle hands, staring out on a back green the size of a table-cloth, surrounded by grimy walls, with no more interesting spectacle to enliven the scene, than the duels, or duets, of the neighbouring cats. So it was, " Helen, I want you to run up this," or " to tack that together," or " just to unpick the other thing," and she became a valuable auxiliary to Plunkett the lady's-maid, not merely with her needle alone,—she soon learned to be very handy with a box-iron !

Of course she was never expected to accompany the family, when they went out in the brougham ; her aunt saying to her in her suavest tone, " You see, dear, your mourning is so recent " (her father was five months dead), " I am sure you would rather stay at home." Accordingly the three ladies packed themselves into the carriage most afternoons, and went for an airing, leaving their poor relation with strict injunctions to " keep up the drawing-room fire," and "to see that tea was ready to the moment of five." Sometimes they gave "at homes," the preparations for which, were left to Helen, who worked like a slavey. These "at homes" were chiefly remarkable for a profusion of flowers, weak tea, weaker music, and a crush.

Next to the cook, Helen was decidedly the most useful member of the household, she was kept fully occupied all day long, and in constant employment, was her only escape from her own thoughts. She was not happy; nay, many a night she cried herself to sleep; her aunt was cool and distant, as though she had displeased her in some way; but to Helen's knowledge, she had given her no cause of offence since the terrible incident of the tea-cup, years and years previously.

Her cousins were sharp, critical, and patronizing, and evidently considered that she occupied a very much lower social status than themselves.

She was unwelcome, an interloper, and felt it keenly. More than once she tried to

screw up her courage, and ask her aunt
what was to be her future.   Undoubtedly she
was not to remain on permanently, as an
inmate of No. 15, Cream Street.   Her big
box still stood in the back hall.   Somehow
she rarely had a chance of a few words with
her aunt alone, her affairs were never once
touched upon in her hearing, and yet she had
reason to believe, that certain animated and
rather shrill conversations, that she frequently
interrupted,—and that fell away into an
awkward silence as she entered a room,—were
about her, and her future destination !

.   .   .   .

Visitors came rapping at No. 15, Cream
Street, every afternoon, and two, out of the
dozens who had called, asked for " Miss
Denis."   A few days after her arrival, she
had been in the drawing-room with her
cousins Carrie and Clara, when her first
caller made her appearance.

The drawing-room was an apartment that
seemed to be all mirrors, lows chairs, small
tables, and plush photo frames—a pretty
room, entirely got up for show, not use.

Several of the chairs, were not to be trusted,
and one or two tables were decidedly danger-
ous, but the *tout ensemble* through coloured
blinds, was everything that was smart and
fashionable, and "good style"—the fetish
the Miss Platts worshipped.

On this particular afternoon Carrie was
yawning over the fire, Clara was looking
out of the window, commenting on a coro-
netted carriage and superb pair of steppers,
with what is called extravagant action, which
had just stopped opposite. Mentally she
was thinking, how much she would like to
see this equipage in waiting at their own
door, when a very curious turn-out came
lumbering along, and actually drew up at
No. 15. A shapeless, weather-beaten, yellow
brougham, drawn by a fat plough-horse, and
driven by a coachman in keeping with his
steed—a man with a long· beard, a rusty hat
(that an Andamanese would have scorned),
and a horse-sheet round his knees.

Little did Helen Denis dream that she
was gazing at that oft-vaunted vehicle—
Lady Grubb's carriage.

"Good gracious, Carrie, who on earth is this?" cried Clara, turning to her sister, who was now staring exhaustingly at her own reflection in the chimney-glass. "And coming to call here! Oh, for mercy's sake, do come and look!"

The door of the brougham was slowly opened, and a very stout old lady, attired in a long black satin cloak, and gorgeous bonnet with nodding plumes, descended, and waddled up the steps.

In the vacant carriage there still remained two fat pugs, a worked cushion, a pile of books, and what certainly looked like a basket of vegetables!

"It's no one *we* know," said Clara contemptuously. "It may be a friend of Plunkett's, or a mistake."

Apparently it was neither, for at this moment the door was flung open, and—

"Lady Grubb!" was announced.

Very eagerly she advanced to Clara, with round, smiling face, and out-stretched hands, saying,—

"So glad to find you at home! My sister

told me to be sure and call, and as I was
at the stores,"—here she paused and
faltered, literally cowed by the expression
of Miss Platt's eyes—Miss Platt, who drew
back, elongated her neck, and looked insolent
interrogation.

" I think you have been so good as to come
and see me," murmured Helen, hastily ad-
vancing to the rescue. " You are Mrs.
Creery's sister ? "

" Yes, and of course you are Miss Denis,"
seizing her out-stretched hand as if it were
a life-belt, for poor Lady Grubb was completely thrown off her balance, by the stern
demeanour of the other damsel.

Helen led her to a sofa, and tried to
engage her in friendly conversation, but it
was not easy to converse, with her two
cousins sitting rigidly by, as if they were on
a board of examination, and not suffering a
word or look to escape them. They sat and
gazed at Lady Grubb in quite a combined and
systematic manner ; to them she was such a
unique object, and such utterly " awful
style."

She, like her sister, was endowed with a copious flow of language, but the very fountain of her speech was frozen by these two ice maidens. The first few words she did manage to utter were hurried and incoherent, but presently she found courage to inquire after Maria, and Nip, and Creery (horrible to relate, she called him " Creery "), and also after many people, she had heard about at Port Blair.

It was very plain to Helen, that Maria had painted her island home, with an unsparing supply of gorgeous colours, and Lady Grubb looked upon her absent relative's position, as something between that of the Queen of Sheba, and the Princess Badoura without doubt. She then murmured a few words of really kind condolence to Helen, and if she had taken her departure at this point, all would have been well ; but she was now becoming habituated to the stony stare of the Misses Platt, and felt more emboldened to converse,—and some malicious elf put it into her head to say, with a meaning smile,—

"I am quite up in all the Port Blair news and Port Blair secrets, you know. I've heard a great deal about a certain gentleman."

Helen became what is known as "all colours," and her two cousins "all ears;" to them she had positively denied that she had left the ghost of an admirer to lament her departure from the Andaman Islands.

"Oh, you know who I *mean*, I can see," continued the old lady playfully. "She had any number of offers," addressing herself rather triumphantly to the Miss Platts, "but Mr. Quentin is to be the happy man," and here the wretched old woman actually winked at Clara and Caroline.

"Indeed, indeed, Lady Grubb, you are quite mistaken!" cried Helen hastily. "Mr. Quentin is nothing to me but a mere acquaintance, and as to anything else, Mrs. Creery—was—was joking!"

"Oh, well, well, we won't say a word about it now, but you must come and spend a long day with me soon and tell me *everything!* I feel as if I know you quite well, having heard of you so often from Maria. I'll just leave my card for your aunt, and now I must really be going," standing up as she spoke. "I suppose Scully is waiting," presumably the uncouth coachman.

The Miss Platts did not ring the bell, neither did they deign to rise from their chairs, but merely closed their eyes at their visitor, as she made a kind of "shy," intended for a curtsey, and wishing them "good afternoon" departed with considerable precipitation.

Helen went downstairs, and conducted Lady Grubb to the hall-door, and presently saw her bowled away in her yellow chariot, with a brace of pugs in her lap.

She was not a very distinguished person certainly, but she meant to be friendly, to be kind, and a little of these commodities went a long way with her now. She blushed when she recalled her cousins' deportment. Surely an Andamanese female, in her own premises, (were they hole or tree), would have shown more civility to a stranger. As she entered the drawing-room, the Miss Platts exclaimed in one breath,—

" What a creature ! Who is she ? "

" She looks like an old cook ! " supplemented Carrie. " I was *trembling* lest any
of our friends should come in."

" Her name is Grubb, she is sister to Mrs.
Creery, the—'' (how could she give any
approximate idea of that lady's pomp ?) " the
principal lady at the Andamans ! " she added
rather faintly.

" Principal lady ! What rubbish ! " cried
Clara. " If she resembles her distinguished
sister, I make you my compliments, as the
French say, on the class of society you
enjoyed out there ! "

" Let us see where she lives. Where's
her card ? What is her name ?—Tubb—
Grubb ? " said Carrie. " Here it is," taking
it up between two supercilious fingers, and
reading,—

𝕷𝖆𝖉𝖞 𝕲𝖗𝖚𝖇𝖇,
Smithson Villas, Pimlico.

" Pimlico ! *So* I should have imagined,"

for, of course, any one who lived in that region was in the Miss Platts' opinion socially extinct.

"You certainly cannot do yourself the pleasure of spending a long and happy day at Smithson Villas," said Carrie with decision. "Goodness knows whom you might meet; and she would be bragging to her cronies that you were *our* cousin."

"I shall go if she asks me," replied Helen quietly. "It is no matter who *I* meet, and I will guarantee, that your name does not transpire."

Was the girl trying to be sarcastic? Carrie looked at her sharply, but Helen's face was immovable.

"Well, I do most devoutly trust that she will not see fit to wait upon you again, or that, if she does, she will come in the laundry-cart!"

"I wonder what the Courtney-Howards thought of her. I'm sure I saw Evelyn at the window," remarked Clara. "Oh!" she added with great animation, "here is the Jenkins' carriage—Flo and her mother.

What a mercy that they did not come five minutes ago!"

Now ensued general arranging of hair, of chairs, and of blinds; evidently the Jenkins were people worth cultivating, and indisputably of " good style."

" Fly away, Helen, at once," cried Carrie, " and tell Price to bring up tea in about ten minutes; and if there is time, you might just run round the corner, and get half a dozen of those nice little Scotch cakes—I know Price hates being sent on messages in the afternoon, and you don't mind."

CHAPTER XIV.

IN WHICH EVERYTHING IS SETTLED TO MRS. PLATT'S SATISFACTION.

" When true hearts lie withered,
    And fond ones are flown,
Oh ! who would inhabit
    This bleak world alone ? "

Moore.

LADY GRUBB'S visit was succeeded by one from Mrs. Home,—a kind, well-meaning little lady, as we know, but as yet attired in what had been a very nice Dirzee-made garment at Port Blair, and even passed muster for best on board ship, but which stamped her at once in the eyes of the Miss Platts as " bad style."

Her boys, too, so eager was she to see Helen, were not yet equipped in their new suits, and were anomalous spectacles in Highland kilts and sailor hats.

Clara and Carrie did not condescend to appear on this occasion, they saw amply sufficient, of Mrs. Home and family, over the dining-room blind.

Helen felt a sense of burning humiliation and shame, to think that now, when she was at home among her own people, they would not even take the trouble to come upstairs and thank Mrs. Home for her great kindness to her, nor even so much as send her a cup of tea. She hoped in her heart that her friend would think they were *out !* But they went audibly up and down stairs, and laughed and shut doors, and Mrs. Home was neither deaf nor stupid.

She stayed an hour, and Helen enjoyed her visit greatly (despite her disappointment at the non-appearance of her relations, or, failing them, the tea-tray). It was one little oasis in the desert of her now dreary life; they conversed eagerly together, and talked the shibboleth of people who have the same friends, in the same country; they kissed, and cried a little, and parted with mutual promises of many letters, for Mrs. Home was going to Jersey, and thence to the Continent.

"Your friends are not our friends, and our friends are not your friends," said Carrie forcibly, and Helen felt that, indeed, as far as appearance went, her visitors had not been a success, and for her own part never dreamt of being admitted within the sacred circle of her cousins' acquaintance.

Now and then she met people accidentally in the hall, or in the street, when walking with her cousins, and once she overheard Carrie saying to Clara, *apropos* of visitors,—

"Of course there is no occasion to introduce Helen to any one," and this amiable injunction was obeyed to the letter. However, the omission sat very lightly, on the once admired of all admirers, at Port Blair.

One morning it happened that Helen was in the drawing-room, when a bosom friend of Carrie's came to call—a Miss Fowler Sharpe, a fashionable and important acquaintance, whom the Misses Platt toadied, for she had the *entrée* to circles barred to them, and they hoped to use her as a pass key.

They made a great deal of this lady, flattered her, caressed her, and ran after her, all of which was agreeable to Miss Sharpe. She was a very elegantly dressed London girl, who spoke with a drawl, and gave one the idea that her eye-lids were too heavy for her eyes. She had come over to Cream Street, to make some arrangements about an opera-box,—and to have a little genteel gossip.

Helen was busily engaged in sewing Madras muslin, and coloured bows on the backs of some of the chairs, where she was " discovered " by her cousins and their friend, to whom she was presented in a hasty, off-hand manner, which plainly said, " You need not notice her ! "

Miss Sharpe stared for a second, vouchsafed her a little nod, then sat down with her back to Helen, and speedily forgot her existence.

The three friends were soon deep in conversation, whilst she worked steadily on, kneeling at the chair she was dressing, with her face turned away from the company.

Their principal topics were, dress and weddings, weddings and dress, and who was flirting with whom, and what was likely to be a match, and what was not, and who looked lovely in such a gown, and what men were in town.

At length Helen, who had not been attending, caught one syllable that made her start and pause,—and then listen with a heightened colour, and a beating heart.

"Yes, I hear that Gilbert Lisle is actually coming back; he has been away among savages this last time, positively fraternizing with cannibals."

"Gilbert Lisle coming home!" cried Carrie. "Then Kate Calderwood will be happy at last.—I suppose it will be all arranged this season?"

"Yes, his father is most anxious that he should settle; indeed, I believe he wrote him out a furious letter, and said that if he did not come home without delay, he would marry again *himself!*" At this threat all three ladies laughed immoderately.

"Imagine any sane woman marrying such an old Turk as Lord Lingard!" drawled Miss Sharpe. "He is seventy, if he is a day, bald and beaky, and with a temper that has a European notoriety; the very idea of his supposing that he would get *any one* to take him!"

"Yes, hideous old creature!" chimed in Clara, "he always reminds me of a white cockatoo with a pink bill."

(Nevertheless, any one of these young ladies would have said "Yes" with pleasure, had Lord Lingard asked them to be his.)

"I cannot imagine how any one ever married him originally," pursued Miss Sharpe; "and yet they say that Lady Lingard was one of the handsomest women of her day."

"Oh, but," put in Clara,—delighted to impart this class of information, "you know, they say she married him out of pique, and she did not live long. I suppose he worried her into her grave."

"No," rejoined Miss Sharpe, "though he *may* have helped to kill her,—she died of consumption."

"Did she? and her eldest son is following her. He is in a rapid decline," added Carrie.

" And you say that Gilbert ·Lisle is really coming home ?" suddenly falling back on the original topic.

" So I'm told. Mother is going to send him a card for our dance ; but I never believe in him till I see him."

" How I wish we knew him ! " ejaculated Clara, looking at her visitor wistfully.

" Oh, you know he is not a society man, only goes to a few houses and some country places where there is good shooting ; now and then you see him at a ball, or in a squash in some staircase ; but he has a very fair idea of his own value, and never makes himself *cheap*," and Miss Sharpe smiled rather disagreeably.

" That's the way with all these rich bachelors ! " exclaimed Carrie. " They are so spoilt, and so abominably conceited."

" I wonder how he got on among the savages ? " said Miss Sharpe.

Little did she guess that the girl who was sitting in the background, with bent head and burning face, could have answered her question then and there.

" I wonder if it will come off with Katie, after all ? " exclaimed Carrie. " She is the girl he used to ride with in the park last year, is she not ?—very freckled, with high shoulders. She comes to our church. I wonder what he sees in her ? " she added.

" It is his father, my dear, who sees *everything* in her : her property ' marches,' as they call it, with the Lingard estates."

" And so she is to be Mrs. Gilbert Lisle ? "

" I believe so." And with this remark the subject dropped.

Helen had listened to this conversation with crimson face and throbbing heart. Everything was accounted for now ; he had been simply amusing himself with her. This man, who was accustomed to be made much of by London beauties, who was eagerly sought for by house parties in country houses, was it likely that he would be really serious in making love to an obscure girl like herself, a girl whom he had come across in his wanderings among savage islands ? " No," she told herself, " not at all likely ; his actions spoke for him. He had been simply seeing

Q 2

how much she would believe, repeating a *rôle* that he had doubtless played dozens of times previously. And during his wanderings, his wealthy destined bride, Miss Calderwood, was all the time awaiting him in England.—*She* was to be Mrs. Gilbert Lisle."

" I do declare, you have stitched that on, the wrong side out ! What can you *have* been thinking of ? " demanded Clara very sharply, when her fashionable friend had departed. " You will have to rip it, and put it on properly.—Your wits must have been wool-gathering ! "

If Clara had known where her cousin's thoughts had been, she would have been very much surprised for once in her life, and ejaculated her favourite exclamation,— " Fancy, just fancy ! "—with unusual animation.

The day after this visit Helen was asked to accompany her cousin Carrie on foot to Bond Street, not an unusual honour. She was useful for carrying small parcels ; true her mourning was shabby,—but none of the Platts' acquaintances knew who she was, and, if the worst came to the worst, she might pass as a superior-looking lady's-maid. On their way back from the shops, Carrie took it into her head, to take a turn in the park. It was about twelve o'clock, and the Row was gay with a fashionable throng of pedestrians. Carrie met several friends, to whom she gave a bow here, and a nod there, and Helen, to her great amazement, recognized one while yet afar off, and although garbed in a frock coat, and tall hat. Yes, she actually beheld Mr. Quentin coming towards her, walking with a very well-dressed woman, and followed by two red dachshunds. She was positive that the recognition was mutual, and was pleased in her present barren life to hail any acquaintance from Port Blair,—even him! When they came almost face to face, she bowed and smiled, and would have stopped, but he merely glanced at her as if she were some most casual acquaintance, swept off his hat and passed on. Evidently Port Blair, and Rotten Row, were two very different places.

A flood of scarlet rushed over her face,

which her quick-eyed companion did not fail to notice, and said,—

" Who was that gentleman ? "

" A Mr. Quentin. I knew him at Port Blair."

" Fancy! I have heard of him—he is quite in society—he is a friend of the Sharpes. I believe he is rather fascinating—but frightfully in debt."

Helen made no reply, but walked on in silence, and Miss Platt put two and two together with much satisfaction to herself. Helen's undoubted confusion, signified, of course, that she cherished an unrequited attachment for this good-looking, faithless man who had just now gone by with a cool ceremonious bow. So much for her cousin's admirers in the Andamans !

.   .   .   .

It was now the end of May, and Helen had been six weeks in London, but so far not a word had been mooted to her about her future plans. She made herself useful, working, shopping, going messages; her aunt admitted to herself, that she was quite

as good as another servant in the house
(though she did not actually use the word
servant, even in her thoughts); she was a
handy, useful, industrious girl, and did not
put herself forward ; so the matter of getting
her a situation had been allowed to remain
somewhat in abeyance.

Helen knew that she must eventually
"move on," but had a nervous dread of
broaching the subject to her relations. Day
after day she failed to bring her courage to
the sticking-point, but the question, ever
trembling on her lips, at last found utter-
ance, and finding herself alone with Mrs.
Platt one morning, she said timidly,—

"Have you made any plans about *me*,
Aunt Julia?"

"Yes, my dear," was the surprisingly
prompt answer, " it is all quite settled ; I had
intended speaking to you before, but something put it out of my head. I have an important letter to write just now, but when
the girls go out this evening, you and I will
have a talk together."

In due time the Miss Platts departed in

the brougham, bound for a little dinner and the play.

Helen, who had assisted to adorn them, partook of a meat tea with her aunt, and then they both adjourned to the little den upon the stairs. There, by the light of a crimson-shaded lamp, Mrs. Platt read the day's news, and Helen sewed, and waited,—waited for a very long time, and, needless to say, she was most impatient to learn her fate.

Her aunt was a lady who never worked, and rarely opened a book, but devoted her whole time to writing, talking, organizing, eating, sleeping, and dressing. She perused the paper as a daily duty, just to see what was going on; and after she had now read every word of it, including advertisements, she folded it up with a crackling noise, and said rather suddenly,—

" This is a capital opportunity for us to have a nice little chat. I have been intending to speak to you for some time; of course you know, dear, that your father left his affairs in a terrible state. I was not the least surprised to hear it, and all that can be scraped together for you is fourteen pounds a year, less than a kitchen-maid's wages," shrugging her shoulders. "There is no use in saying anything about the dead; what is done, is done; nor that, to satisfy his ridiculous ideas of honour, he left his only child—"

"No, no use, Aunt Julia, for I would not listen to you," interrupted Helen with sudden fire. Mrs. Platt was astounded; this outbreak recalled old days, she positively recoiled before the expression of her niece's eyes, the imperious gesture of her hand. She leant back in her chair with folded arms, and sat for some moments in indignant silence, when she reached out two fingers and pulled the lamp-shade down, so that her face was completely in the shadow. She had reason to do so, for she was going to say things of which she might unquestionably be ashamed; and once more she commenced, as if repeating something she had previously rehearsed,—

"Ours is the oddest family, we have so few relations on the Denis side, no nice connections, no influential friends ; when your grandfather " (why could she not say my father ?) " came to such a fearful smash, all his old associates abandoned him, as rats leave a sinking ship. I married, and made new ties, your father married too ; but, as far as I know, your mother had no respectable belongings ; my sister Christina also made a wretched match,—she married a half-crazy Irish professor she picked up at Bonn—he afterwards came in, for some miserable Irish property on which he lives, but *he* could do nothing, he can hardly keep the wolf and the bailiffs from the door as it is ; Christina, as I suppose you know, died last Christmas."

"No, Aunt Julia, I never heard of it."

"Oh, well, of course it does not affect you." (Nor did it apparently much affect Mrs. Platt.) "She and I, had not met for many years. Then there is my Aunt Sophia—your grand-aunt. She is an invalid, and lives at Bournemouth, scarcely ever leaving her room. She is very wealthy, and we correspond constantly, but most of her money goes to charities, in which she takes an interest,

and unfortunately she takes no interest in
you. She has got it into her head that you
are worldly!"

Helen stared round the lamp-shade, to see
if her aunt was joking.

"It's quite true," responded Mrs. Platt,
meeting her gaze, " and once she gets an idea
into her head,—there it stays! So it is rather
unfortunate; but, at any rate, all her thoughts
are at present centred on a mission to the
Laps. Then," with a perceptible pause, "we
come to myself. I am not a rich woman"
(though she strained every nerve to appear
so, and had upwards of three thousand a
year). "I spend every penny of my income,
and am often pressed for money. Of course
in the country, or at the seaside, we would
have a margin, but the girls would not hear
of living anywhere but in town,—and naturally
I have to study them, and their interests."

" Of course, Aunt Julia," acquiesced her
listener.

"This is a ruinous neighbourhood, and
this house, though so tiny, costs four hundred
a year; no doubt for half that sum, I would

get a mansion in Bayswater; but, as the girls
say, there is no use in being in town at *all*, if
you don't live in the best part of it, and here
we are! Then we require to keep up a
certain style to correspond with the situation,
—a man-servant is indispensable, and a carriage; the horses, of course, are jobbed.
Again, we have to entertain, to go to the
seaside, to dress—and this last, even with
Plunkett making half the things, costs a
small fortune! The long and the short of it
is that, out of my very tolerable income, I
never have a single sixpence at the end of the
year. This being the case, you will readily
understand, my dear Helen, that, much as I
should wish to do so, I cannot offer you a
home here."

"No, of course, Aunt Julia, I never expected you to do so," replied her niece in a
low voice.

"You are a sensible girl, wonderfully so
for your age, and I talk to you, you see, as
openly and frankly as if you were my own
contemporary! I could not afford to dress
you as you would require to be dressed, and

to take you out; besides, the brougham is a crush for three, as it is, and three girls at a dance would be out of the question. I must say, I should have liked to have given you a season, but, as Clara points out, my taking you into society, would entail leaving one of them behind, and charity begins at home; and, candidly, I am very anxious to see them settled."

"Yes, aunt, of course I understand that your own daughters should come first."

"And besides all this, my love," waxing more affectionate as she proceeded, "I really have no room to give you. Plunkett requires one to herself; there is mine, and the girls', and the spare room, and, you see—"

"I see, Aunt Julia," interrupted her niece, "don't say another word. And now what are your plans for me?"

"Well, I had hoped to have got you a very happy, comfortable home, with a very rich old lady in the country, who required a nice cheerful young girl to talk to her, and read to her, and be with her constantly. She was rather astray mentally—a little weak, you know; but you would have got two hundred a year. However—" and she stopped.

" However, aunt—?"

" Well, I heard indirectly, that she was liable to rather *violent* paroxysms occasionally, and came to the conclusion, that it would not do ! I have been making inquiries among my friends—of course, it's rather a delicate business, and I don't mention that you are my own niece ; it would be so very awkward, you know ; but I hope to hear of something suitable ere long. Meanwhile, dear, I'm sure you won't be offended at my telling you, that we shall want your room next week ! "

Helen's hands shook, her lips trembled, so that for the moment she was unable to speak. Was she to be turned out of doors ? She had exactly four pounds in her purse upstairs !

" Clara's rich godmother always comes to us for June," continued Mrs. Platt, " and we have to study her, and to make the house bright and pleasant ; it is then, we always give our little dinner parties. We do our

best to please her ; she is very liberal to
the girls, and we could not possibly put
her off. She will have the spare room, as
usual,—and her maid always occupies *yours*."

" Yes, Aunt Julia."

" I have made a very nice, temporary,
arrangement for you, dearest ! A lady I
know, who keeps a large school at Kensing-
ton, has most kindly offered to take you
gratis for a month or two,—till we can look
about us. You are to teach the younger
classes French and music."

" In short, go to her as governess ? "

" Oh, dear me, no," irritably ; " it is a mere
friendly offer. She obliges you, you oblige
her, as one of her staff has gone home ill,
and she is rather short-handed just now."

" And will she pay me ? " inquired Helen
as bluntly as Mrs. Creery herself.

" Oh, no, I don't think there was any
reference to that ! Perhaps your laundress
may be included ; but you scarcely seem to
understand that she is going to give you
board, and lodging for *nothing*. You are not
sufficiently experienced for a governess ! "

"But—" began Helen, thinking of her superior musical talents and fluent French.

"But," interrupted her aunt tartly, "if you can think of any other expedient for a couple of months, or have a better suggestion to make, let us have it, by all means!"

Her hearer pondered. There was Miss Twigg, Miss Twigg no longer; she was married, and had gone out to Canada. Mrs. Home was in Germany, her former schoolfellows were scattered,—to whom could she turn?

"Of course this is a mere temporary step, as I said before," urged her aunt. "I shall do much better for you in the autumn; I have great hopes of getting you a comfortable home through some of my friends, and as a favour to *me*. So, meanwhile, will you go to Miss Gibson's or not?"

"Yes, aunt; I will do whatever you please."

"Very well, then, that is settled. I must get your things done up a little first. Your Aunt Sophia sent ten pounds for you, and I was thinking, that as the girls were going out of mourning—three months, you know, is ample for an uncle—that you might help Plunkett, to remodel one or two of their dresses for yourself."

Helen felt a lump in her throat, that nearly choked her. She would wear a cast-off garment of Mrs. Home's with pleasure, and accept it as it was meant; but Clara's and Carrie's!—never! And she managed to stammer out,—

" No, thank you, Aunt Julia; I shall do very well."

" But that black every-day dress, is not fit to be seen."

" It will do in the schoolroom,—and I shall get another."

" Now I consider that wanton extravagance; when you can have Clara's for nothing. Perhaps your dignity is offended?" and she laughed at the mere idea of such a possibility, and then added, " By the way, *are* you proud?"

Helen made no reply, but bent her eyes on her work.

" Then, my dear child, the sooner you get

rid of that folly the better,—for poverty, and pride, are no match for one another."

"How soon did you say I was to go to Miss Gibson's, aunt?"

"On Monday next. You can leave your big box here still, and if you like to come over to lunch every second Sunday, you may do so. But I doubt if you will care for the long walk across the park,—or if Miss Gibson could spare a servant to walk home with you."

"Then, thank you, I won't mind."

"Well, dear," rising as if a load had been removed from her mind, "I believe we have settled everything satisfactorily. It is so much pleasanter to talk over these matters face to face. And now, love, I'll say good-night. I daresay you would like to finish Carrie's handkerchief before you go up-stairs." Then, stooping and kissing her, she added, "Be sure you put the lamp out carefully," and with this parting injunction, Aunt Julia opened the door, and departed, leaving her orphan niece alone with her own thoughts.

Helen stitched away mechanically for nearly ten minutes, then she laid down her work, and sat with her hands lying idly in her lap, and her eyes riveted upon the rose-coloured lamp-shade, but her thoughts did not take any reflection from that brilliant hue. The life that had begun so brightly, now stretched out before her mental vision as grey and dreary as a winter's day. She was imperiously summoned to work for herself, to take up her post in the battle of existence, to toil for her daily bread for the future,—her only aim being to lay by some provision for her old age; she saw before her years of drudgery, with but this end in view. She had no friends, no relations, no money. A cold, dull despair settled down upon her soul, as she sat in the same attitude for fully an hour. At last she rose, folded up her work, carefully extinguished the lamp, and then made her way noiselessly up to her own apartment under the slates.

 END OF VOL. II.

LONDON:
PRINTED BY GILBERT AND RIVINGTON, LIMITED,
ST. JOHN'S SQUARE.